THE AUTOMATION
LEGAL REFERENCE

A guide to legal risk in the automation,
robotics and process industries

THE
AUTOMATION
LEGAL REFERENCE

A guide to legal risk in the automation,
robotics and process industries

Mark Voigtmann

Illustrated by Aaron Reiter

Notice
The information presented in this publication is for the general education of the reader. Because neither the author(s) nor the publisher has any control over the use of the information by the reader, both the author(s) and the publisher disclaim any and all liability of any kind arising out of such use. The reader is expected to exercise sound professional judgment in using any of the information presented in a particular application.

Additionally, neither the author(s) nor the publisher has investigated or considered the effect of any patents on the ability of the reader to use any of the information in a particular application. The reader is responsible for reviewing any possible patents that may affect any particular use of the information presented.

Any references to commercial products in the work are cited as examples only. Neither the author(s) nor the publisher endorses any referenced commercial product. Any trademarks or tradenames referenced belong to the respective owner of the mark or name. Neither the author(s) nor the publisher makes any representation regarding the availability of any referenced commercial product at any time. The manufacturer's instructions on use of any commercial product must be followed at all times, even if in conflict with the information in this publication.

Copyright © 2013 International Society of Automation (ISA)

All rights reserved.

Printed in the United States of America.
10 9 8 7 6 5 4 3 2

ISBN: 978-0-876640-08-1

No part of this work may be reproduced, stored in a retrieval system, or transmitted in any form or by any means, electronic, mechanical, photocopying, recording or otherwise, without the prior written permission of the publisher.

ISA
67 Alexander Drive
P.O. Box 12277
Research Triangle Park, NC 27709

Library of Congress Cataloging-in-Publication Data in process

This book is intended as a general guide to legal risk in the automation realm and should not be relied upon as advice for any particular situation. The law applicable to any given circumstance can be nuanced. Please consult counsel.

Table of Contents

List of Figures   ix

Preface   xi

Chapter 1   Automation Projects and Legal Risk   1

Chapter 2   Project Delivery Methods   9

Chapter 3   Proposals and Purchase Orders   13
    TIP: Beautiful Proposals and Toxic Purchase Orders   16

Chapter 4   Scope of Work   19
    CHECKLIST: Five Reasons to Watch Out for "Incorporated" Contracts   22
    CHECKLIST: Five Ways of Knowing the Real Scope of Work   23

Chapter 5   The "Dirty Dozen" Contract Clauses   25
    CHECKLIST: The Three Most Important "Missing" Contract Terms   31
    CHECKLIST: 10 Reasons Not to Agree to Indemnify   32

Chapter 6   The Other "Ugly Eight" Contract Clauses   33
    CHECKLIST: Five Things You Should Know about Warranties   37
    TIP: Should FAT and SAT Be Mentioned in Contract Documentation?   38

Chapter 7   Negotiating Automation Contracts   39
    TIP: Is a Letter of Intent Binding?   42
    THE CSIA RIDER   43

Chapter 8   Specifications   47
    TIP: The Power of Performance Specs   49
    TIP: What If the Specs or Plans Are Defective?   50
    TIP: The Pitfalls of Owner-Specified Equipment   51

Chapter 9   Intellectual Property   53
    CHECKLIST: Six Questions to Ask when Setting Automation License Royalties   58

Chapter 10   Automation Standards   59

Chapter 11   Professional Licensing   65
    CHECKLIST: Nine Realities of Professional Licensing for Automation Companies   67
    TIP: How Licensing Works in the U.S.   68
    TIP: How Licensing Works Internationally   68

Chapter 12 "Green" Considerations   69
  CHECKLIST: Six Reasons Automation Companies Need to Speak Green   74

Chapter 13 Changes and Other Mid-Project Communications   75
  TIP: Paying Attention to Fine Print at the End of a Project   78
  TIP: Managing Changes on Automation Projects   79
  TIP: Do Too Many Changes Open Up an Entire Automation Project to Renegotiation?   80

Chapter 14 Dispute Resolution   81
  CHECKLIST: Four Types of Proceedings   90
  CHECKLIST: Three Indispensable Inquiries to Make Before Litigating   91
  CHECKLIST: Eight Rules for Resolving Contract Disputes   92
  TIP: Beware of Making False Claims   93

Chapter 15 Negligence   95
  CHECKLIST: Three Critical Ways to "Manage" Negligence   99
  TIP: Who Is Responsible for Functional Safety?   100

Chapter 16 Insurance   103
  CHECKLIST: Nine Additional Insurance Coverages and Concepts Worth Knowing   109

Chapter 17 Liens, Bonds and Other Remedies   113
  CHECKLIST: Six Remedies Other than Liens or Bonds   120

Chapter 18 Maintenance and Service Agreements   121
  TIP: Making Clear that Perfection Is Not Possible   125

Chapter 19 Legalities for Tough Economic Times   127
  CHECKLIST: Four Legal "Gut Checks" for Automation Providers during Tough Times   129

Chapter 20 Auditing Legal Health   131

Chapter 21 Working with Attorneys   135
  CHECKLIST: 10 Additional Tips for Getting More Value Out of Your Lawyer   137

Glossary   139

Index   155

About the Author   159

List of Figures

Figure 1—Design-Bid-Build and Design-Build. 11

Figure 2—Determining the Scope of Work for a Project Can Be Like Opening Nesting Dolls. 21

Figure 3—The Dirty Dozen Contract Clauses. 25

Figure 4—The Ugly Eight Contract Clauses. 34

Figure 5—Frequency of Words (by Size) in Responses to Informal Automation Standards Survey. 60

Figure 7—Categorization of Some Automation Standards. 63

Figure 6—The Spectrum of Automation Standards. 63

Figure 8—Three Types of "Green" Certification. 70

Figure 9—The American Civil Discovery Process. 83

Figure 10—The Key Ingredients in Litigation Soup. 84

Figure 11—Types of Motions in American Civil Litigation. 87

Figure 12—Liability for Negligence and Breach of Contract Can Overlap. 96

Figure 13—The "Umbrella" of Insurance Coverage. 105

Figure 14—The Logic of the Subcontractor Exception. 107

Figure 15—A Two-Company Payment Dispute. 113

Figure 16—A Two-Company Payment Dispute with a Lien Claim Added. 115

Figure 17—A Two-Company Payment Dispute with a Payment Bond Claim Added. 118

Figure 18—Strengths and Weaknesses of Maintenance Agreement Types from Automation Provider Perspective. 122

Figure 19—Response Boundaries According to Severity Level. 125

Figure 20—The "Healthy Company" Documents. 133

Preface

This book springs from years of writing and speaking about legal risk affecting automation projects. Most of the subject matters were first explored in articles written for industry publications (most notably in my "Legalities" column in *Control Engineering* magazine), in speeches before organizations (most frequently the Control System Integrators Association, but also the International Society of Automation and World Batch Forum) and in newsletters written for clients.

I have a passion for trying to explain complex topics in simple terms. This perhaps is a reflection of my decade as a journalist before becoming a lawyer, but it is also a necessity in the automation world. Here, lack of communication is a two-way problem. Not only do most lawyers have little or no understanding of what automation companies do (mention the word "software" and many will politely look for the exit), but I also have seen all too many engineers (and, frankly, automation company executives) attempt to navigate their way through legal hazards in what can only be described as a "penny wise but pound foolish" manner—saving a few thousand on legal fees on the front end only to see a company-killing problem arise as a result of that inattention.

This book is for both audiences. For automation industry insiders, my hope is that the writing is clear enough to permit the understanding of some very important risk management concepts without the inconvenience of going to law school. For lawyers with clients in the automation industry, I am hopeful this book has value as a sort of "checklist of intersections" between the parallel universes of automation and law (although lawyers will quickly note this does not remotely resemble a legal treatise—I met my goal of not using a single legal citation).

It could be argued that I come to the automation field as a matter of pedigree. My father spent his career as an engineer, then plant manager, then executive for Eaton Corporation, and as a child, I spent many hours exploring the Eaton manufacturing facility in Batavia, Illinois. Back in those days, and even much later (I worked my first summer job at another Eaton plant close by), I gave no thought to how such facilities and the systems within them came to be constructed, let alone all of the things that could go wrong.

Of course, those were relatively simple places compared to today's automated facilities. As best as I can recall, the childhood plants I visited back then consisted only of a series of workers arrayed at workbenches operating independent machines. The automated systems I write about in this book are both more and less than that—*more*, in that they are infinitely more efficient and productive (sometimes with robots working tirelessly in the place of people); *less*, in that they are also infinitely more fragile due to the fact that any one integrated component can effectively shut down many others.

Given that fragility, the legal risks awaiting those who venture out in the automation world are by no means small. Although reading this book will not eliminate those risks, I hope it will at least demystify them—so that each new project is begun with open eyes.

Chapter 1

Automation Projects and Legal Risk

At first glance, the legal risks facing the automation industry would appear to defy categorization. Are such projects about instrumentation? Are they about services? The answers to these questions are varied. One common way to speak of automation projects is to say that they involve "delivery" of equipment; another is to characterize them as the "sale" of a product. Yet another is to speak in terms of the "installation" of industrial computers or the "development" of software. While all these classifications are more or less accurate (not to mention applied with some frequency), none of these fully captures the essence of this type of endeavor—and therefore none points to the optimum means of managing legal risk.

The best way to approach risk management in automation is to treat such projects as a very specialized type of *construction project*. Not only does this approach permit examining these projects in a somewhat more dynamic and sequential (as opposed to static and scattered) way, it makes it possible to apply a very useful and well-established vocabulary—even if it is a vocabulary more traditionally reserved for projects involving pouring concrete and erecting curtain wall.

Yet there may be something of an industry bias against this approach. Even in the most greenfield of projects, it is not unusual to see a bright line imposed between the construction side of the project and the MRO (maintenance, repair and operations) side—often including parallel lines of authority and completely different personnel. Leaving aside for a moment the incongruity of the "M" and "R" parts of the MRO acronym in the greenfield context (i.e., because the maintenance and repair activities in a greenfield project would seem to be few in number), the split does provoke a question that is worth asking: are the risks that much different? The answer is "No." The risks in both are imposed by contract.

Speaking the Language of Contract

Realizing that automation projects must navigate the world of construction contracts does not exactly require a leap of faith. So why do automation practitioners frequently deny this fact? One reason is the lingo. "We don't have a contract—we just have this purchase order," is what more than one control system integrator has told me. Yet as any first-year law student quickly learns, an exchange of proposal and purchase order is not the opposite of a contract—it is just another *type* of contract. In other words, it is a member of the same family as a thirty-page, single-spaced document with formal signature blocks on the last page.

But here is perhaps the most difficult part to understand: very few automation companies recognize the fact that the legal expense of unraveling a dispute involving the type of contract reflected in a simple purchase order is often much greater than for a contract with 30 pages of lawyer-speak. Why is such a dispute more expensive? The reason is so obvious that it tends to be completely overlooked. In the proposal and purchase order realm, the legal relationships, obligations and rights of both parties are much less clearly defined. So when a dispute erupts, you will be paying your lawyer to assert what you *meant* to say instead of what you *did* say. That takes more time—and time, of course, is money.

The good news is that legal risk in automation projects can be successfully managed by using the language of contracts at the front end. Now for the bad news: there are two distinct vocabularies (let's call them "dialects") in this area that must be mastered to maximize project success.

The Dialect of the Traditional Construction Contract

The first dialect to be mastered is that of the traditional construction contract. This dialect—which speaks of change orders and indemnity and "liquidated damages"—is important for two reasons. First, it is the dialect nearly everyone else on the project will be speaking (and therefore it is an advisable way of communicating with those persons). Second, it is a very old dialect, with a great many useful terms with application to automation projects. Among the traditional construction contract terms of maximum importance (each of which will be explored in greater detail in the coming chapters of this book) are the following:

- **Integration, Incorporation by Reference, Order of Priority and Dragnet Clauses.** These terms refer to one of the most fundamental—and (ironically) neglected—questions of all. What scope of work was agreed upon? What happens when one part of the contract disagrees with another? What happens when agreement on a particular point is unclear?

- **Change Order Clauses.** What are the options when scope is arguably added to or subtracted from the project? Can these changes be unilaterally imposed?

- **Payment and Pay-When-Paid Clauses.** How does payment work? What conditions must be met for payment to be made? Does it matter if others are not paid?

- **Liquidated Damages, Force Majeure and No-Damages for Delay Clauses.** What are the risks and options when the project takes longer than expected? Does it matter whether the delay was extraordinary or routine? What if there is a suspension of all activity? Does it matter whether it was an "act of God" or that someone was responsible?

- **Limitation of Liability and Consequential Damages Clauses.** If something on the project goes wrong, are there limits to the potential legal liabilities? Could one of the participants held responsible for the lost profits of others?

- **Indemnity Clauses.** Did any of the participants agree to extend a "protective shield" over the project in any respect? What problems does this shield protect against?

- **Warranties.** What is the duration and scope of any warranty? Are warranties implied by the law but not stated in any written part of the agreement? (Yes, you heard me correctly—agreements implied by circumstances, but not written down.)

- **Claim, Mechanic's Lien, Bond, Retainage and Dispute Resolution Clauses.** How are disputes to be resolved? Are liens an option or is the project bonded (or both)? How are retainage amounts (withheld funds) released—and can claims be made upon the retainage of another? How and where must disputes be resolved?

Knowledge of all these terms—and the contract paragraphs in which they are defined—goes a long way toward successfully managing risk in an automation project. Still, if automation risk management were a course of study, mastering these terms would yield only a middling grade, because you must also know the language of risk that is particular to these projects and few others—namely, the dialect of automation.

The Dialect of the Automation Contract

The exercise of automation law requires new words and new approaches (beyond those supplied in the traditional construction project) in at least three areas. Those areas are changes, performance and intellectual property.

Changes. It has been said that construction contracts are more challenging than other types of contracts because, unlike the other contracts, construction contracts are being written at the same time as they are being performed. While this is a bit of an exaggeration, the underlying point is worth bearing in mind. It is simply an acknowledgement that modifications of contract terms—most often in the form of agreed-upon changes or as a result of unforeseen events—are not the exception to the rule, but are the rule. Indeed, it would be a most unusual construction project (perhaps there are none) in which the original specifications, terms and conditions remained unchanged through final completion.

Take that reality, multiply it times ten, and you will have a sense for the multiplicity of changes that automation-related companies (whether end user, system integrator or engineering firm) typically confront. By their very nature, automation projects are constantly changing. For maximum success in managing automation project risk, therefore, you must find a way to manage this anticipated, but frustrating state of affairs.

The flood of changes can be managed in at least two ways. The first way is to control the definition of "change." Because the final scope of the software component, for instance, of an automation project is an inherently mobile target, it is in the interest of owners and end users to limit the definition of "change" to a certain threshold or criteria. Meanwhile, it is in the interest of control system integrators and engineering firms to avoid any such limitation, while carving out a point at which excessive changes are deemed a breach of the contract (in traditional construction parlance, this is known as a "cardinal change").

The second way is to control how adjustments due to change are *processed*. As in the first case, owners and end users will want to impose strict notice criteria and tight timeframes for seeking adjustments. Integrators and engineering firms, on the other hand, tend to benefit from relaxed or nonexistent notice provisions and deadlines.

Because the occurrence of changes is so predictable on automation projects, the importance of addressing changes at the front end cannot be overemphasized.

Performance. Automation contracts are also different from traditional construction projects because they almost always involve the construction of a system with moving parts. For that reason, specifications often do not measure performance by static criteria (e.g., was this thing built in conformity with these particular dimensions and consisting of these specified materials?) Instead, the specifications have a tendency to be either output-based (does the thing that was built produce 100 widgets per hour?) or they are satisfaction-based (is this thing that was built fit for its intended purpose?) The first type of specification is called a "design spec," and the second two are variations on what is called a "performance spec."

The tug-of-war between owners and automation contractors is also (predictably) present here. Naturally, owners want to be satisfied, and should push for the most subjective performance spec that is possible. Automation contractors, on the other hand—especially if they are fulfilling a design that is largely dictated by an owner's consultant—will want to limit their construction commitment to building that which was specified without any guarantee as to performance. Prevailing in this struggle is not easy. It can literally mean the difference between a project that is labeled by the contract terms as a success or a failure.

Intellectual Property. Because automation is a subspecies in the proliferating realm of computer technology, the automation industry is naturally evolving at a faster rate than the traditional world of "sticks and bricks." New ways of solving problems do not appear in fits and starts, but in nearly every project. Indeed, one way of thinking about the automation industry is as a type of engine for solving problems—one that routinely asks, for example, "How might we best control this process?" The result is often not just an elegant solution, but a valuable one.

But who owns the solution? Is it a "work for hire," paid for by the owner or end user and therefore entirely within its ownership and control? Or is it a valuable invention of the system integrator that must be retained because it is vital to its future success as a business? What happens to the underlying legacy intellectual property of an integrator that may be incorporated into the design? What about commercial off-the-shelf software that is included? Are there limitations to the future uses of the application—or have the parties decided to license the technology in some way between them?

Obviously, the answers to these questions are enormously important, not just for project success, but for long-term *enterprise* success on the part of both system integrator and owner.

Knowing the Neighborhood

While it may be all well and good to know (and to speak) both the traditional construction contracting dialect and the more specialized automation one, such fluency only goes so far when you are operating in an atmosphere in which people would prefer not to engage you in a conversation.

You: "Let's talk about terms."

Them: "Sorry, we are not permitted to change these terms."

Welcome to the world of the take-it-or-leave-it contract.

Both owners and automation companies may have good reasons for this attitude. First, it is, without question, easier. Saying "we are not permitted to change these terms" beats the back-and-forth of negotiations (and, heaven forbid, dealing with lawyers) every time. Second, market leverage or business conditions ("you need us more than we need you") may make taking such a position all too tempting—the contractual equivalent of "shooting fish in a barrel." Third, it minimizes risk for the companies that successfully impose their own terms.

Naturally, there are tactics for dealing with this approach if your company happens to be on the receiving end. Among them: bidding with exceptions, providing balanced (i.e., non-threatening) alternate terms, appealing on the basis of fairness, making the argument that risk should reside with the company with the most control over that risk, or even—when all else fails—incorporation of a proposal, rider or other document into the larger contract that may round off the sharpest edges. (We will explore each of those tactics in more detail later in this book.)

But even if there is no negotiating—not even a conversation—that does not mean that knowing the language of contracts is without value. There are always two decisions that, regardless of negotiating stance, cannot be taken away from any party in an automation project: the decision on price and the decision as to whether to participate in the project in the first place.

That is why it is a central argument of this book that the contract dialects must be learned—and that attention to these dialects must be paid. Whether it involves establishing a process for non-lawyer company managers to analyze the "paper" of a deal or involves hiring lawyers to do so is a matter of ability and preference. The important thing is that someone performs a focused review.

Chapter 2

Project Delivery Methods

If you were mingling at a social gathering and—forgive me for this—found yourself cornered by someone else connected to the automation industry (I know, it sounds like a rather boring party), how would you casually describe the project you were about to embark upon?

You might say (using the shorthand of fellow automation people): "We're migrating from an X to a Y platform."

Or you might say: "We are replacing legacy infrastructure with fieldbus technology."

Each of these is an attempt to begin defining the playing field. Yet in the world of legal risk, such language does not say very much. That is because these descriptions are entirely technology-based; the important legal relationships that are being carved out, knowingly or unknowingly, are left unstated.

Now, if you *are* interested in giving attention to that topic (and, of course, you *should* be interested, if controlling legal risk is anywhere on your radar screen), it is important to fit the upcoming project into the right legal category. The million-dollar, front-end question is literally this: What is the project delivery method?

Although "project delivery method" can mean a number of things, a useful definition for our purposes is to think of it as a decision to allocate legal responsibility in a particular way among a given set of project participants. In automation terms, think of the project delivery method as the operating system, with the contracts or purchase orders being executable software. Each such "operating system" carries with it advantages and disadvantages, and a unique set of risks for the automation project.

Most Common Project Delivery Methods

Design-Build. Many, and perhaps most, automation projects are "design-build." At its core, this project delivery method simply means that the same outfit that is designing the automation system is also installing it. (You may know this method by at least two other names: turn-key or EPC [engineer-procure-construct]—although the meanings are not exactly the same). If, as the system provider, you identify the project as design-build, your company has effectively bitten off two very large mouthfuls of risk: the design risk ("Was this thing designed correctly?" "Is the HMI user-friendly?" "Did we adequately assess the end user's performance needs?") as well as the construction risk ("Did we do a professional job?" "Did we follow the manufacturer's recommendations?" "Did we install this right?").

For the purchaser of automation hardware and functionality, the "turn-key" label is perhaps the most appropriate term, because it reflects the essential advantage of the design-build method. What the owner or end user has purchased is the ability to "turn the key" and have the system work. Or, to use a consumer term, it is "one-stop shopping." When there are problems (and this is only in theory), there can be no finger-pointing because both design and construction are delivered by the same player.

Design-Bid-Build. Add an outside designer (engineer) to the mix and you have what lawyers refer to as "design-bid-build." Here, the design is first negotiated between owner and engineer; then the builder (in our world, the automation provider) bids on that job. Although this is the most prevalent system in the realm of traditional sticks and bricks construction, it is less used on automation projects for the simple reason that system design and execution can be difficult to separate. The indispensable (and frequent) exception to this tendency toward the unification of automation design and construction involves what I would refer to as the "uber-designer"/sub-designer split in larger projects, in which the layout and conceptual engineering for an entire range of systems is worked out in advance—with the automation provider's solution for a particular application constrained by the predetermined framework (often with particular platforms and devices already specified).

Chapter 2 Project Delivery Methods

Figure 1—Design-Bid-Build (left) and Design-Build (right).
The difference to the owner is the number of parties.

Although these are the "big two" project delivery methods applicable to automation projects, there are others:

Construction Management. Like using an uber-designer in the design-bid-build project delivery method, this arrangement adds an "uber-contractor" to coordinate installation, which is common in projects where there is competition for access to work areas and numerous interdependencies among deliverables.

Design-Build-Operate-Maintain. This is the ultimate in turn-key arrangements. Not only does the automation provider agree to design and install a system that will do what the owner requires, the provider agrees to remain on site beyond commissioning—and (for an additional fee, of course) run and maintain the system for an extended period of time (typically training the owner's forces in the process).

Integrated Project Delivery. This project delivery method has received a lot of attention recently because it is built on a radical model. Take a large project with multiple designers and contractors, then put them in the same room at the earliest possible stage to collectively design, manage and construct a set of systems—even going to the extent of signing a single contract. Sounds great, doesn't it? Well, before you begin singing "Kumbaya," try to wrap your mind around exactly how this would work. If you are having difficulty doing that, you are recognizing the essential weakness of integrated project delivery—

which is that the devil can be in the details. Nonetheless, the idea is certainly intriguing and the gains (on paper, at least) are potentially enormous.

The Link Between Method and Risk

Now, why should you go through the exercise of identifying which project delivery method your project fits into? There are at least four reasons:

First, it is vital to match the appropriate project delivery method to the owner or end user's goals. It can mean the difference between the success and the failure of a project.

Second, it forces you to reflect on important risk-related fundamentals that are, or should be, in the contract: Are we asking this provider to design anything? Is design split between two entities? Who is coordinating installation? Do the warranties match the contractual obligations?

Third, it impacts insurance requirements. From the owner or end user's perspective, for example, professional liability insurance covering the engineer can be essential when you add one part design deliverable and one part smaller engineering firm.

Fourth, before you can cut down a tree, you need to find the forest—it's that simple. The forest is the project delivery system; the tree is the risk.

Chapter 3

Proposals and Purchase Orders

I was speaking at an automation conference a couple of years ago when an executive came up to me and said, "You can talk about contracts all you want, but what you're saying simply doesn't apply to many of these companies. Some of these people operate on a handshake. The rest get by with proposals and purchase orders."

Part of what he said was true. The integration of a facility is often not the subject of a detailed written contract with, say, 30 single-spaced pages (and four or five exhibits).

But his other point—that contract law has little application to the automation environment—now, that was just plain wrong. Here's why: Handshakes and purchase orders, not to mention 30-page, single-spaced agreements, are all contracts. They all determine the rights of the parties involved. They can all be legally binding. And they all have the ability to turn an otherwise profitable business opportunity into a nightmare if the wrong terms are included.

Notch up "we don't use contracts" as one of the top misconceptions of automation companies when it comes to legal matters. Here are three others that quickly come to mind: competing terms, contract form incompatibility and intellectual property traps.

Competing Terms

Perhaps you feel legally protected as long as you transmit a copy of your terms and conditions sheet with your proposal or purchase order. But are you really? For instance, what happens when your customer (or integrator) responds by sending you its own contradictory terms and conditions? Does it mean the deal is off? Probably not—especially if the participants go forward with the project in spite of it (as they typically do). But that exchange of "dueling T's and C's" may change the very nature of the project you are undertaking and dramatically alter the extent of the legal protections you thought you had.

How? Try this on for size. It is sometimes the case (at least in the United States, according to a law on the books in nearly every state called the Uniform Commercial Code) that conflicting terms cancel each other out, and only the remaining agreed upon terms are recognized as binding. Or, if you are doing business in the European Union, one or more of your terms may simply be stamped unenforceable by the government in question.

There can be ways of getting around these unintended results through careful drafting, but the operative word is "careful." Paying careful attention to the details is far and away the most important message of this book. Failure to do so is like playing a game without knowing the rules.

Contract Form Incompatibility

If the process or control piece is just one small component of a much larger construction project (as it often is), you may find yourself being asked to sign your name to a contract that looks and sounds very much like a construction contract—for instance, the A401 subcontractor agreement put out by the American Institute of Architects (AIA). Other common AIA forms include the A101 and A102 agreements (for prime contractor), the A107 (for projects of limited scope), the A141 and A142 agreements (for design-build projects) and last, but not least, the A201 (the "general conditions" that are incorporated into many of the other forms). But do such forms work for construction involving automation companies—or is it like trying to insert the proverbial square peg into a round hole? For the following reasons, I vote for the latter:

First, the AIA forms do not adequately address commissioning (unless your idea of commissioning is a punch list).

Second, the AIA warranties are going to be difficult to translate to the world of sensors and internal relational databases. For example: Is software from third parties being warranted? How is the warranty affected if the information supplied by the end user is deficient?

Third, the intellectual property (IP) provisions are almost certain to be insufficiently detailed for the needs of the end user, integrator or equipment supplier. (For more on these and other incompatibilities, see the checklist at the end of this chapter.)

Intellectual Property Traps

While we are on the IP topic, many companies make the mistake of simply taking IP for granted (making the assumptions that either "I own this" or "I will own this"). But the risks are many. Perhaps you've heard of the patent infringement lawsuits filed a number of years ago by a company called Solaia Technology against many users of control systems. The patent in question (U.S. Patent No. 5,038,318) related to PLCs that fed data directly into a spreadsheet program, thereby allowing operators to view the operation of the PLCs with a personal computer interface. Regardless of what you may think of such lawsuits, defending against them is costly. And the issue of who pays the attorneys might have been addressed in the automation contract at the front end—whether in the form of a purchase order or detailed agreement.

TIP: Beautiful Proposals and Toxic Purchase Orders

An integrator's proposal can be a beautiful thing. In some cases it is a lot like reading the history of the company, with a three-color logo at the top, high-minded statements about the integrator's corporate philosophy in the first paragraph, and carefully-worded descriptions in the middle about how the integrator adds value to its industry, along with reassuring phrases such as "state of the art" and "we stand behind what we do."

The best of these proposals, of course, also contain well thought-out and appropriately protective terms and conditions that make clear exactly what the integrator has agreed to do and *not* agreed to do and that also limit the integrator's liability to a fair amount. What's not to feel good about? So you, the integrator, go ahead and sign the customer's purchase order and do the project without objection.

Take a moment and pat yourself on the back.

But what if the purchase order that you (the integrator) just signed is what I call a "toxic purchase order"? If that is the case, you might as well take your beautifully-written proposal and flush it down the pipes. That is because the purchase order cleverly killed the applicability of your proposal—in favor of the PO's own terms and conditions.

A toxic purchase order (or contract) makes clear the following:

- The terms in the PO are the only terms that matter
- Any other terms and conditions are expressly rejected
- The scope of work is described in the RFQ or the customer's specs

Either your proposal is called out as inapplicable or it is not even mentioned!

There are variations, of course. One version of the toxic PO does at least go to the trouble of mentioning your proposal, and maybe even including it as an official "contract document." But then it makes it utterly clear that if there is any inconsistency between the PO and your proposal, the PO is supreme.

Now there can also be a positive effect of such "toxic" purchase orders. It may be the case that your proposal contains such great marketing language that it could come back to haunt you later—and the proposal-killing PO has just saved your bacon. For instance, there are numerous court decisions about how marketing phrases like "state of the art" can create a *warranty* to that effect for the buyer.

So it may be that when the customer's PO comes back looking toxic, with a flat statement saying that the only terms that matter are the ones in the PO, it may be doing you, the integrator, a big favor (literally by saving you from yourself).

CHECKLIST: Six Reasons That Well-known Contract Forms May Not Be Appropriate

A number of organizations sell contract forms that are generally used in the construction industry. Among the most well-known of these are those published by the American Institute of Architects (AIA), Associated General Contractors (AGC) and the Engineers Joint Contract Documents Committee (EJCDC). Are these adequate for automation work? There are at least six ways in which they may not be:

- ✔ The forms do not adequately focus on the sometimes-shifting interplay of legal responsibilities between facility designer, system designer, system installer and equipment manufacturer.
- ✔ Intellectual property concerns are often inadequately addressed.
- ✔ The warranties are inappropriate for software-related deliverables.
- ✔ The concepts of acceptance testing, commissioning and maintenance are typically missing.
- ✔ Such agreements are usually silent on the issues of existing mechanical equipment and/or legacy automation systems.
- ✔ Rigid provisions establishing the scope of work and the procedure for making changes cannot always be reconciled with the necessarily fluid nature of some automation projects.

Chapter 4

Scope of Work

The phrase "scope of work" refers to the boundaries of the project. Inside the scope of work is each and every item of work that was agreed to be performed. Outside the scope of work is everything else. The gray area, if there is one—and I have rarely seen an automation project where there is none—is where the disputes arise. One side says (just as an example) that checking the compatibility of existing mechanical equipment with new software was part of the work to be performed. The other side might say (again, just as an example), "We took the mechanicals as we found them—and if they are inadequate, it cannot be on us." Who is right? Write a check to your lawyer for many thousands of dollars and I guarantee that you will eventually get an answer from a judge, jury or arbitrator. (It may not be what you want to hear, but there will be an answer!)

What Is the Scope?

This is obviously a situation to be avoided, but exactly how might that be done? Clearly, a close examination of the "specs" is part of the drill. (By "specs," I mean anything in writing that purports to describe the work being done, which can include the proposal, purchase order, contract, addenda, drawings or formal specifications labeled as such.) Are there any ambiguities? Have all potential contingencies been addressed? Most process automation companies are quite capable of running through such a checklist from the technical angle, but there is one place where an important safeguard is often left out—checking whether the contract is "integrated."

Say the word "integration" to a control engineer and you may quickly find yourself transported to the realm of DCS and SCADA. But contract integration can be equally important to the success or failure of a project. Take it from a lawyer who spends considerable time navigating within the automation world: It is amazing how many companies neglect this most basic of legal principles—

to the point where the *system* may be integrated but the *legal requirements* are not.

Contract Integration Traps

There are two typical ways in which contract integration can go off track. The first is the lack of an integration clause in the project documents. This is the paragraph of the contract where it says (and I am paraphrasing): "This document that we have just signed is the only document that matters. None of the other documents that we previously exchanged or oral promises that we previously made mean a darn thing. In fact, please feel free to take any other documents (including both the RFP and the accepted proposal) to your backyard and burn them."

In case you are wondering, such a paragraph (even as overblown as I have made it) is binding throughout North America. But the larger point is that if there is *not* such a paragraph, it can be very unclear which terms apply to the project and which do not.

The second way for a contract's integration to unravel is through what lawyers call "incorporation by reference." You know those Russian nesting dolls where one painted wooden doll can be pulled apart to reveal another figure of the same sort inside—which itself can then be pulled apart to reveal yet another—and so on? That image probably is a pretty decent stand-in for this legal concept.

Let's say that before you sign a contract or a purchase order, you check it quickly for any problematic terms. Nothing jumps out at first glance, except that there does seem to be a list of items identified as "other contract documents." You pass that by and sign.

Guess what? You may think you have reviewed the whole agreement, but you have not. Every one of those other listed items—each likely with its own "nested" terms—is also a part of what governs your company's performance. Before you sign, better get your hands on each and every one of them and figure out how they relate. Otherwise, you may be on the hook for obligations that you have never even talked about—let alone reviewed ahead of time in written form. (In what is perhaps the most common version of this, a related contract to which your company is not a party is incorporated by reference, including all of its specs.)

Figure 2 shows the multiplicity of ways in which a group of documents can be interrelated to form a contract that bears little resemblance to the scope of work originally contemplated.

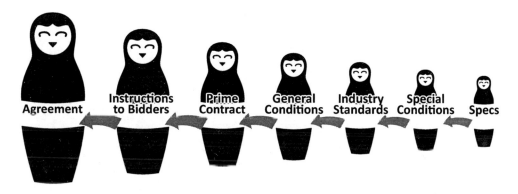

Figure 2—Determining the Scope of Work for a Project Can Be Like Opening Nesting Dolls.

Now for a surprise ending: There are those situations (quite a few of them actually) where a lack of contract integration—in the legal sense of the term—can be advantageous. The essential question to ask is, "In whose interest is clarity—ours, theirs, or everyone's?" While I would argue that in most cases the answer is "everyone's," there are exceptions. And if that is the lay of the land, you are hereby granted permission to go forth and make the contract less integrated—and intentionally more confusing.

CHECKLIST: Five Reasons to Watch Out for "Incorporated" Contracts

Subcontracts involving the automation or technology piece of a project often incorporate the terms of an upper tier contract to which the automation company is not a party. These "incorporated contracts" are far from benign. Parties to such subcontracts should be especially wary of these provisions because of the following concerns:

- ✔ **Acting in the Dark.** It is one thing to incorporate a contract to which your company is not a party. It is another thing to actually track the language of that separate contract and take its additional risks into consideration. Operating out of two contracts can be unwieldy.

- ✔ **Notices of Change.** Parties to an automation subcontract are often required to give written notice of a change, and the timing and content of such a notice may depend on what is in the incorporated contract.

- ✔ **Dispute Resolution.** Parties to an automation subcontract may be bound to resolve disputes by whatever means is set forth in the incorporated contract, which may require, for example, arbitration.

- ✔ **Limitation on Damages.** In many subcontracts, an automation subcontractor's damages may be limited in certain circumstances to whatever the contractor can recover. In such instances, parties should be especially aware of any damage limitations contained in the incorporated contract.

- ✔ **Warranties.** Automation subcontractors should determine whether any warranties related to their work are part of the incorporated contract.

CHECKLIST: Five Ways of Knowing the Real Scope of Work

- ✔ **Was there a signed contract?** If not, the scope of your agreement may be subject to state commercial codes, which make assumptions that do not always line up with what was intended.

- ✔ **Is there an "integration" clause in the contract?** This is the part of the contract that tells whether it is "open-" or "closed-ended." If closed-ended, preliminary project documents—such as an RFQ or integrator proposal—may no longer be a part of the final deal.

- ✔ **What documents are incorporated in the contract?** Whether closed- or open-ended, additional documents listed as "contract documents" in the central document must be reviewed in detail for a true appreciation of the risks involved.

- ✔ **What standards are incorporated in the contract?** The same goes for published standards, regulations or industry protocols. Even the single use of an abbreviation may carry significant meaning—and thus expand the scope of work.

- ✔ **Is there a dragnet clause?** Dragnet clauses are indefinite but potentially problematic contract clauses, which purport to require greater scope whenever there is alleged ambiguity as to how much work is required.

Chapter 5

The "Dirty Dozen" Contract Clauses

Want a good starting place for figuring out whether to accept another company's terms and conditions? Start by looking for the "Dirty Dozen" contract clauses depicted in Figure 3. Each of these 12 clauses defines a legal battleground of prime importance for companies engaged in automation projects—whether user, integrator, engineer or OEM.

Whether any of these is truly "dirty," of course, is a matter of perspective, or to put it another way, one company's "dirty" is another company's "being careful." But by going through the project documents to check the treatment given to each of these topics, you will gain an appreciation for whether the project (contractually at least) conceals time bombs or is reasonably fair.

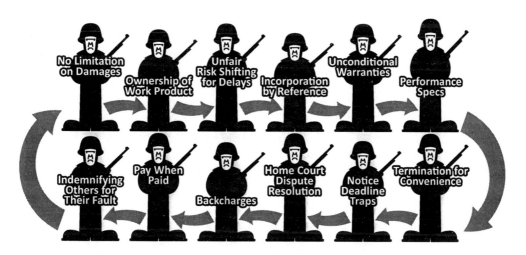

Figure 3—The Dirty Dozen Contract Clauses.

No Limitation on Damages

Contract clauses addressing limitations on damages are so tremendously important that they arguably should occupy several spots on the list. The two key phrases here are "consequential damages" and "damages cap."

Consequential damages are lost revenues and other downstream losses that might be suffered from a facility being shut down or a software solution being defective. From the perspective of integrators and other service providers (which tend to be smaller companies), the prospect of being held responsible for the six and seven figure losses that could conceivably result from a manufacturing facility's shutdown is the equivalent of betting the business on a single project. It's utterly unacceptable and unfair—and, surprisingly, accepted by service providers without objection every day of the week.

Those vendors who know better give careful attention to requesting a "waiver of consequential damages." Although this will do nothing to avoid the cost of repairing or replacing a system that fails (so-called "direct damages"), it eliminates the company-wrecking devastation that would result if the company was forced to reimburse the end user for downstream production losses.

Now, from the process owner's perspective, the picture is quite different. Why shouldn't a control system integrator be liable for lost profits if its system does not make the grade? For that reason, some companies, including at least one well-known automaker and one heavy machinery manufacturer, make it a policy never to waive the right to collect consequential damages in the event of an automation system failure.

Is there a middle ground? There is, but it takes some flexibility (and creativity) on the part of both owner and vendor. Among the most common methods are the following:

- Making consequential damages collectable up to the amount of applicable insurance coverage.

- Capping the amount of consequential damages collectable under any circumstances to a percentage (or sometimes a multiple) of all amounts paid under the applicable contract.

Ownership of Work Product

If the contract describes software or another deliverable as a work for hire, the creator of that software is giving up its intellectual property rights. It goes without saying that this does not make sense for a company that is marketing that creation (in one form or another) to multiple users in the technology marketplace.

A better solution (from the perspective of the creator at least) may be granting a non-exclusive, non-transferable license to the invention—assuming, of course, that it has the bargaining power to demand it.

Naturally, the purchaser of the deliverable has a different view. It is buying a bundle of technology and it almost certainly has several providers to choose from. Why should it not be entitled to all of the fruits of that purchase?

Both sides can be right, of course—and here too there may be a middle ground to be had if such is the case. One approach is to distinguish base or legacy intellectual property (the inventions that form the core of the seller's business) from the intellectual property that was created specifically for a customer. But the dividing line is not always easy.

Unfair Risk Shifting for Delays

Contract forms sometimes present an automation company with a two-edged sword: provisions imposing harsh damages for delays attributable to its failures (these are called liquidated damages or sometimes simply LDs), but an absence of similar remedies if the company is held up because third parties are not prepared to proceed or there are other circumstances beyond the company's control. Unfortunately, these one-sided "no damages for delay" clauses, even if unfair, are enforced by the laws of most states.

While such clauses may seem to be simply the problem of providers of panels, transmitters, cable trays and related services, (and not the concern of the purchasers of those services and equipment who are typically the beneficiaries of that type of unfair risk sharing), this can be a shortsighted view. For one thing, there are sometimes hidden costs to unrestrained one-sidedness. Among those costs: (a) compensating claim tactics on the part of service providers ("We'll pick a fight with you because the best defense is a good offense"), (b) a dissatisfied vendor marketplace without the possibility of repeat business due to lack of profitability, and/or (c) higher prices.

Incorporation of Other Contracts by Reference

Contracts that make reference to other documents that are not immediately available, but seek to "incorporate by reference" those documents (thus making them a binding part of the deal) can be dangerous for obvious reasons. From the service provider's perspective, this practice can be problematic because the actual scope of work may be different from what the core contract or even the specifications may suggest. From the process owner's perspective, an unrestrained series of incorporations by reference may produce ambiguities that lead to conflict—and disputes. (For a more detailed discussion, see Chapter 4.)

Unconditional Warranties

Development of a control system requires coordination and communication between the end user and integrator. Only sophisticated end users realize the degree to which their participation is critical to the usefulness and functionality of the end product. From the perspective of the integrator, warranties should be linked to cooperation from the end user and should not be unconditional. End users, on the other hand, should strive to make warranties as open-ended as possible and should ensure that clauses requiring participation in troubleshooting are not disruptive to the business.

Performance Specifications

Do the specs require the automation company to build something according to a particular plan, or do they require it to build something that has a particular functionality (e.g., "process a hundred widgets per hour")? The latter are called performance specifications and should be avoided by automation companies and, if unambiguous, embraced by end users. (For more on this topic, see Chapter 8.)

Termination for Convenience

It is becoming more common for owners to insist that their contracts include a clause which gives them the right to terminate a contract "for convenience," which is to say "for no good reason at all." For an owner, this can be a necessity due to the changing priorities of upper management. For an integrator, this is not fair treatment because landing a project can prevent taking on other work.

Notice Deadline Traps

Look carefully at notice deadlines. Forty-eight hours for giving notice of unforeseen conditions, changes, or a claim can be impossible to meet. If you cannot change them, at least highlight them so your people know that the clock is ticking. (Even if the deadline is impossible to meet, giving notice as soon as possible may be legally sufficient in many circumstances.)

Home Court Advantage Dispute Resolution

Does the contract require all disputes to be heard in a courtroom in the home county of the end user, five hundred miles from the project site and three states away from you?

Backcharges

Does the contract give the customer virtually unlimited opportunities to assess backcharges? This is a desirable right in an end-user-dominated world (and some would argue this is the one we live in). From the perspective of a service provider, however, such clauses have the potential of being oppressively applied. If fairness is the objective, there should be a notice deadline beyond which these charges are deemed waived. (And there should also be a process for documenting them.)

Pay When Paid

On its face, this is strictly a concern of a service provider or a middleman contractor. If a subcontractor vendor's contract is not directly with the end user, there may be a clause specifying that payment is not due unless the middleman has been paid by the end user. The middleman wants this clause for an obvious reason—cash flow. But a downstream subcontractor may find itself penalized for a reason that has nothing to do with its own performance. For instance, the middleman might not have been paid due to its own deficiencies or that of an unrelated subcontractor. (Thus, the subcontracting vendor in that situation should argue that such a clause may be effective only to the extent that its work is somehow linked to the problem causing the nonpayment.)

Even though this interaction is taking place a rung or two below the end users, they too may have an interest in the fairness of such clauses. Some U.S. states permit filing a mechanic's lien by an unpaid subcontractor against the real estate of the owner, notwithstanding the fact that a pay-when-paid clause in the subcontractor's contract makes clear that payment is not yet due.

Indemnifying Others for Their Own Fault

Generally speaking, indemnification means protecting someone from the claims of a third person. Unfortunately, it is very common to see contract forms that require vendors to indemnify a customer for losses caused by the customer's own negligence. This makes the vendor, in essence, the insurance company for the customer! (This is one reason not to agree to such a clause. For a long list of various things that can go wrong with indemnification clauses, see the end of this chapter.)

* * *

Does the absence of the "Dirty Dozen" (however the word "dirty" is defined) mean that a contract can be signed without risk? Absolutely not. Nonetheless, it does go a considerable distance toward ensuring that the terms and conditions are fair—or, at a minimum, that the possibility of disputes has been lessened.

Chapter 5 The "Dirty Dozen" Contract Clauses

CHECKLIST: The Three Most Important "Missing" Contract Terms

If your contract is missing the terms below, you are exposing your company to significant risk.

- ✔ **Limitation of liability.** Not putting in place a limitation of liability could break your company (for instance, in the event your workers cause an uninsured loss with significant revenue impacts). Ideally, there should be a hard cap on the total dollars for which your company may be liable. At the very least, there should be a limitation on "consequential damages" — the most likely type of damages to go into six or seven figures.

- ✔ **Protection of intellectual property.** Give your customers the products and services they are paying for—but do not give away the tools in your tool box. The contract should clearly distinguish between work product developed for the customer and your pre-existing intellectual property. You should also clearly define the parties' rights in connection with each type of intellectual property.

- ✔ **Fair indemnification.** While indemnification typically cannot be avoided, each party should be responsible for third-party losses only to the extent that such losses are caused by its fault and negligence. If the other side refuses to reimburse your company for losses caused by its own fault or negligence, your alarm bell should be sounding.

CHECKLIST: 10 Reasons Not to Agree to Indemnify

✔ The indemnification refers to the performance criteria of a control system.

✔ You are not 100% certain what "indemnify" really means.

✔ The indemnification agreement goes beyond personal injury and property damage—and addresses core business issues.

✔ The indemnification is one-sided (you are indemnifying the other company, but the other party is not indemnifying you).

✔ You are indemnifying the other party for an adverse event, even if the other party is solely negligent.

✔ You are indemnifying the other party for 100% of the adverse consequences of an event without any discount for the other party's contribution.

✔ You are required to "provide a defense."

✔ You have never taken the time to "sync" your insurance program with your contractual obligations.

✔ You are being asked to indemnify parties other than the one with which you are contracting.

✔ Blanket indemnification is being provided against patent infringement without attention to "pass through" patents.

Chapter 6

The Other "Ugly Eight" Contract Clauses

Did you really think I was done? Not by a long shot. The "Dirty Dozen" list in the preceding chapter arguably contains only the most problematic clauses lurking in automation-related contracts.

In this chapter I thought I would add to the list. Call these the "Ugly Eight"—bringing the number of major offenders to an even 20. As with the prior list, whether any of the "Ugly Eight" are truly "ugly" depends on who you are; which is to say, one company's "ugly" is another company's "being careful." (Translation: Put them in your arsenal, Mr. End User. Avoid them, Mr. Integrator.)

"Free from Defects"

Perhaps there are some construction projects—somewhere—where the phrase "free from defects" is appropriate in a contract. But automation projects are not one of them. Control engineering, by its very nature, involves software—and has there ever been a piece of software truly without defects? A global search for the word "defect" should be standard operating procedure in automation contract scrutiny. If not removed entirely, the phrase "free from defects" might be replaced with reference to best practices or industry standards.

Dragnet Clauses

Perhaps you have seen these. Paraphrased, they state—and I am not making this up—"On the off chance there is any ambiguity about what your company agreed to do, you and we agree that you agreed to do whatever is worse for you." Now, depending on exactly what that extra obligation is, there may be arguing room as to whether that is truly a binding commitment. Still, in the slippery, slidey world of automation contracts, who wants to be burdened with trying to extricate himself from such a mud hole?

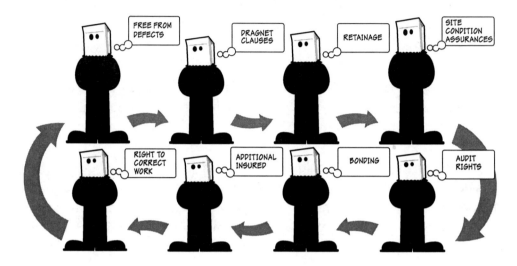

Figure 4—The Ugly Eight Contract Clauses.

Retainage

In plain language, these clauses provide the following: "For every ten dollars your company is owed, we will pay it nine—holding the final dollars till the end." While these "good faith" holdbacks sometimes make sense (e.g., in ensuring solvency or workmanship or payment of subcontractors), there is no reason to blindly apply them. Are there equipment purchases requiring a larger fronting of cash? Is the upstream contractor (if there is one) subject to the same retainage? Is the percentage negotiable? Might it be reduced at the halfway point?

Site Condition Assurances

It is quite common for automation contracts to require a provider to vouch for the claim that the existing conditions of a facility—and sometimes even design components contributed by others(!)—are sufficient for the purposes of the upcoming project. Requiring such assurances can be fair to a limited extent; for instance, it is certainly not unreasonable to expect an integrator to have visited the site (at a minimum). But such assurances can start crossing the line when they have the provider acting, in essence, as an insurance policy against unforeseen conditions or ill-conceived prior design work.

Audit Rights/Financial Condition Assurances

If you are a control system integrator and you have a lump sum contract, does the end user have the right to look at your books? (If so, why? Does that really make any sense?) Conversely, if you are the end user and a provider is working under a cost-plus arrangement, is there any reason you should not have this right? While the proper approach in those situations may be relatively clear, the issue is a bit trickier when it comes to confirmation of overall financial condition. Lump sum or not, if there are signs that a provider is becoming insolvent, it seems reasonable to require some show of financial health mid-project—most would agree. But what degree of assurance is appropriate? The matter is negotiable.

Bonding

Bonds reflect a guarantee from a third party (a surety) that a company will fulfill its obligations (e.g., paying its subs, installing a system that meets contract requirements). Although it is common to pass through the cost of bonds to the end user, procuring them is perhaps the Catch-22 of contracting. At the risk of some overstatement, bonds are arguably only needed when the bonded party's financial stability is somehow questionable. But they can only be procured if that stability is unquestionable! To observe that contracting around bonding is challenging is to state the obvious.

Additional Insured Endorsements

The additional insured endorsement is another typical contractual requirement that can leave automation-related companies in unintended knots. Agreeing to such an endorsement means asking your insurance company for an endorsement that gives someone else the right to make a claim on your insurance policy. The pitfall for the automation contractor is that there is an extra cost to this endorsement—or it may not be available at all (e.g., it's never available on professional liability policies). The pitfall for the owner is that it's not enough to have an additional insured endorsement required by the contract. You must also have the actual endorsement in hand. Owners often neglect that second step.

Right to Correct Work

If the provider does not fulfill its contractual obligations, does the owner have the right to step in to perform the work instead? It sounds like a reasonable remedy, except for the usual collection of variables: Is it just any failure to perform that triggers the right, or only an important failure? What notice, if any, must be given? Following notice, does the provider itself have the opportunity to correct the deficiency? Is the warranty period extended if the fix works? Must notice be given again if the fix does not work? If the owner exercises its right to correct work, are all of the provider's remaining obligations cancelled or terminated? Creating a fair "right to correct" provision can be complicated.

* * *

So there you have them—the other Ugly Eight. They may not make the all-star squad, but they are there on the sidelines ready to spoil your project.

CHECKLIST: Five Things You Should Know about Warranties

✔ **Warranties can be created before you even sign a contract.** Not all warranties are created when a contract is signed. In trying to make a sale, sales or marketing personnel may exaggerate system capabilities or otherwise make promises about an automation system that it may or may not be possible to fulfill. Promises or representations made prior to the parties' actually executing the contract can create *express* warranties (warranties based on the parties' oral or written expressions of agreement).

✔ **Most of the time, warranties can be disclaimed.** Despite representations that might be made during the course of contract negotiations, it is usually possible to disclaim all warranties except for those that become a part of the written contract. But certain requirements must be met for this outcome, particularly with respect to *implied* warranties (those warranties imposed by the law regardless of what the parties agreed).

✔ **The parties can contractually limit liability for breach of warranty.** There are ways to limit liability for breach of warranty, including placing a cap on damages and/or providing an exclusive remedy of repair or replace.

✔ **Beware the "battle of the forms."** Parties will often exchange their own form contracts, and neither party will sign the other party's contract. The rules concerning which terms apply in these circumstances can be complicated at best. But parties to contracts involving automation equipment and services should be aware that these complicated rules may dictate whether certain warranties have been created or effectively disclaimed.

✔ **Disclaimers don't always work.** Just because a contract contains language that disclaims warranties, such disclaimers don't always work, especially if there is contradictory language elsewhere in the contract.

TIP: Should FAT and SAT Be Mentioned in Contract Documentation?

Put aside for the moment the question of whether you typically negotiate detailed written contracts with your customers. (In case you have not gathered this already, it is my view that wherever possible, you should.) A more specific question concerns FAT (factory acceptance testing) and SAT (site acceptance testing). Should there be explicit mention of either one or both in whatever documentation formalizes your project obligations? In other words, should not only the fact that FAT and/or SAT will occur—but also the protocols and standards governing that testing—be written down? I think the answer is an unqualified yes.

Consider some things that can go wrong if FAT and/or SAT requirements are either unclear or left unsaid:

- Disagreement over the definition of a successful test (e.g., "Your facility for the FAT does not approximate the demands of ours.")

- Disagreement over the scope of testing (e.g., "You have not tested the interface with Level 3.")

- Non-participation (and therefore non-sign-off) by key players among the end users where such participation is not required by an agreed protocol. This may create acceptance problems later (e.g., did you involve not just manufacturing and engineering, but also maintenance, IT and quality assurance personnel?).

There are different places where FAT/SAT requirements can be contractually spelled out. One place, naturally enough, is the contract, terms and conditions, or proposal (because FAT and SAT protocols are highly specific to each company, the usefulness of a standardized set of provisions is probably limited). Another place is in the specs. Failing either of these, a decent backup approach is to submit FAT and/or SAT protocols to the owner or design rep during the course of the project and obtain sign-off in advance of the testing (and hopefully in advance of any problems). Of course, the danger in the "after the fact" method is that tendering the document may serve to open up a can of worms.

The essential point with FAT and SAT is that clearly-articulated testing protocols are nearly indispensable to minimizing the legal risks connected to automation projects of every type and size. Indeed, it would not be an overstatement to say that failure to give front-end attention to testing is like beginning a competition without knowing how the judges are grading. (Or, put another way, while a "customer is always right" mentality can carry one far in business, there is nothing wrong with knowing that some favorable contract terms are tucked away in a back pocket for use in a less amicable setting.)

Chapter 7

Negotiating Automation Contracts

One of the frustrations of being a lawyer who advises automation providers is that—sometimes—there is only so much I can do. (Note to reader: this chapter is for integrators without much negotiating leverage. Large end-user companies with all the negotiating leverage can skip to the next chapter.)

Here's an example: Let's say Control Engineers Limited wants legal advice in connection with a new project. Appearing in my inbox is a little gift from Control's customer, consisting of a 40-page set of general conditions, a five-page set of special conditions, some intimidating specs and a purchase order with very fine print.

"This is going to take me a few hours to go through but I can already tell you we are going to need to change a whole lot of this to make it fair," I might say in response.

Why is it that I only "might" say this?

Because long ago I learned my lesson. In many cases, the reply from my client is the following: "Well, unfortunately, the customer says he is following a corporate policy never to change a project's terms and conditions."

What's a lawyer to do with that? (After all, what good is a lawyer who can't take out his red pen to justify his fancy diploma?)

The answer is that lawyers must do what clients have always done: adjust to the realities of the customers, not the other way around. Flexibility is king.

What I mean here is that "take it or leave it" contracts require a flexible strategy. There are at least five potential responses. (By "flexible," I mean that if a strategy fails, you move to the next—although not necessarily in this order.)

Bid with Exceptions

The assumption here is that your company is bidding on a project in which at least some of the customer's terms and conditions are known ahead of time (although that is not absolutely necessary to make this work). In the "bid with exceptions" strategy, you create a simple and generalized list (an exceptions sheet) that specifies the sorts of terms that are not agreed upon—and you append it to your proposal. This strategy has two advantages. First, your company presumably has already shown its technical brilliance in response to the RFQ—maybe all of the "good stuff" in your bid will make the customer overlook the page of exceptions that you have attached. Second, you do not need a lawyer to create an exceptions sheet. Your sheet would simply note items like: "Bid is conditional on working out a fair sharing of design risk." Or even: "Control Engineers Limited assumes no liability for lost profits of the customer." The details can be worked out later.

Provide Specific Alternative Language

If there is a contract term that is particularly worrisome, it can be effective to show the customer the not-so-bad substitute language that you are proposing. Invite the customer to point out any problems with it. The downside of this approach is that some amount of legal fees may need to be incurred to produce a substitute clause that is legally sustainable.

Appeal on the Basis of Fairness

This may seem obvious, but whenever you are able to engage a customer in a focused discussion on the fairness of a particular term, you are already 75 percent of the way home. Indeed, a customer's statement that "we never change our terms" is nothing less than an attempt to avoid this discussion. Talking through the real-world fairness of a particular, problematic clause can serve to make the legal problem less theoretical. One warning: To the extent you are successful in getting fairness on the agenda, you should be careful in picking your targets. There should only be one or two so that you do not wear out your welcome.

Appeal on the Basis of Risk Control

A widely-recognized tenet of good risk allocation says that risk should be allocated to the party that is best situated to affect the outcome. Some types of risk that owners try to pass on to contractors do not meet this test—and can be rebutted accordingly. Once again, if you are successful in provoking a conversation, you have already advanced your cause significantly.

Lawyer Up

Sometimes the open involvement of lawyers, particularly as to disputes, can be counterproductive because it has the undesirable effect of inflaming the situation. This is less often the case with contract negotiations, however, particularly when the automation provider is negotiating with a buyer or other non-lawyer corporate official whose flexibility to negotiate may be hampered by "official corporate policy." "Lawyering up" can break the logjam. The involvement of a lawyer for the automation provider tends to result in the buyer calling upon its own legal department. This can result in greater flexibility on the part of the end user because the lawyer who takes the place of the buyer probably wrote the corporate policy in the first place—and knows where it can be bent and where it cannot.

Incorporate Your Proposal

Getting your proposal listed as one of the official contract documents is a common negotiation strategy. Of course, this assumes that your proposal comes with a set of terms and conditions that are more favorable than the customer's—and it also assumes that you are able to navigate the several types of killer contract clauses in the customer's documents that seek to neutralize the effect of your proposal (for this you may need a lawyer).

* * *

There are other strategies. For instance, some companies have succeeded in using a trade association "rider" (the Control System Integrators Association has a good one—see below) by telling their customers that "My association won't let me sign your agreement unless we also sign this." The advantage of the CSIA Rider is that it also contains non-lawyer text that provides in written form many of the strategies suggested in this chapter. This explanatory text is not just for the guidance of one negotiator; it is intended to be reviewed and considered by both contracting parties.

Of course, if no changes are permitted, the best companies involved in the automation world ultimately go through a formal or informal risk assessment process (with or without a lawyer's assistance) that analyzes whether a deal is really worth it. After all, simply walking away can be the smartest strategy when you are faced with a "take it or leave it" contract.

> ### TIP: Is a Letter of Intent Binding?
>
> Letters of Intent are temporary agreements between parties that are in the process of negotiating a more complete deal. Such letters, by definition, are non-binding in the sense that there is no guarantee the ultimate "full deal" will be agreed upon—but they are binding in two other important respects: (a) they bind the parties to participate in good faith negotiations of a definitive agreement (and create liability if one party or the other does not act in good faith) and (b) they bind the parties to any terms contained in the Letter of Intent itself. An example of such a binding term included in the letter might be the obligation to compensate the other for work performed before a definitive agreement is agreed upon—regardless of whether it ever is. Another example might be an agreement to include a particular term in the eventual definitive agreement. An important warning about letters of intent is that if there is work being performed (in the hope a definitive agreement will be agreed upon) it does function as a sort of temporary contract as to the parameters of liability for that work. So just like every other automation contract, it can be extremely important to pay attention to all the important elements.

THE CSIA RIDER

Printed with the permission of the Control System Integrators Association

This RIDER (the "Rider") amends the contract dated _____ (the "Contract") entered into between _____ ("Customer") and _____ ('Integrator") for _____ (the "Project" or "Work"). The terms "Project" and "Work" have the meanings that they have in the Contract, or, if not defined in the Contract, the meanings that they have in the integration and automation industry.

Paragraph 1: The purpose of Paragraph 1 is to insure that the Integrator and Customer communicate effectively with each other. The CSIA believes that such communication is key to a successful project.

1. **Communication.** Integrator will communicate regularly with Customer regarding progress on the Work. Customer will provide sufficient qualified technical assistance to ensure that the specifications for the Work and the requirements of the Customer are communicated to the Integrator in a timely and understandable manner consistent with the agreed scope of the Work. The parties will take all steps necessary to secure the availability of such technical personnel at appropriate times to coordinate and communicate with each other.

Paragraphs 2 and 3: The design and development of an integrated system is a cooperative process. Paragraph 2 requires the Customer to actively cooperate with the Integrator in this development. In return, the Integrator will provide the warranty in Paragraph 3.

2. **Design, Development and Training.** The agreed process for design, development and training is described in **Attachment A**. Any party that disregards or fails to fully participate in good faith in the design, development or training process described in Attachment A waives and releases the other party from all claims, including, without limitation, liability for negligence, breach of warranty or breach of contract, arising from such party's disregard or failure.

3. **Warranty.** The Integrator warrants the hardware supplied for the Project to the extent of any manufacturer's warranty to Integrator that is applicable to such equipment. The Integrator warrants that it will, for a period of _____, correct, repair or replace software that fails to meet the requirements of the Contract under normal use. This warranty is void if anyone other than Integrator has modified the software.

EXCEPT FOR THESE WARRANTIES, THE INTEGRATOR MAKES NO REPRESENTATION OR WARRANTY OF ANY KIND WHATSOEVER, INCLUDING BUT NOT LIMITED TO, ANY WARRANTY OF FREEDOM FROM PATENT INFRINGEMENT, OF MERCHANTABILITY, OF FITNESS FOR A PARTICULAR PURPOSE, OR ARISING FROM A COURSE OF DEALING OR USAGE OF TRADE OR OTHER EXPRESS OR IMPLIED WARRANTIES.

Paragraph 4 strives to expedite the resolution of disputes in a fair manner outside of a courtroom.

4. **Resolution of Disputes.** If a dispute arises out of or relates to this Contract and if the dispute cannot be settled through negotiation, the dispute shall be decided by arbitration administered by the American Arbitration Association under its Construction Industry Arbitration Rules, and judgment on the award rendered by the arbitrator(s) may be entered in any court having jurisdiction. The law governing this Contract shall be the place of the Project but the place of the arbitration shall be a neutral location that is neither the home city nor state of the Customer or Integrator. The prevailing party in the arbitration shall be entitled to recover its attorneys' fees, interest and arbitration expenses as elements of any award.

Paragraph 5 represents a fair sharing of intellectual property rights. Customer is permitted to use developed software for its desired purposes while permitting the Integrator to retain ownership of the means, methods and products that are essential to its future success as a business.

5. **Intellectual Property.** The software to be developed by the Integrator shall remain the sole intellectual property of the Integrator. Following acceptance and final payment to the Integrator, the Integrator will grant to the Customer a non-transferable, non-exclusive license to use the software for the Customer's internal purposes only.

Paragraph 6 is a recognition of a key principle in this RIDER—that risk and responsibility should be assigned in proportion to each party's ability to control the outcome. It also places a reasonable limit on the liability of both the Customer and Integrator to each other.

6. **Limit of Liability.** The total liability of the parties to each other for any loss, indemnity, damage or delay of any kind except payment of the price will not under any circumstances exceed the lesser of thirty-five (35%) of the Contract Sum or the applicable insurance amounts covering the liability. Under no circumstances will any party be liable to the other for any loss, indemnity, damage or delay arising out of its failure to perform due to causes beyond its reasonable control, including, without limitation, acts of God, interference by others, delays in receiving approvals or necessary information from each other, fires, strikes, floods, war, terrorism, riots, delays in transportation and adverse weather. Under no circumstances will either Integrator or Customer be liable to each other for any special, incidental or consequential damages.

Paragraph 7 keeps the Integrator and Customer from hiring away each other's employees.

7. **No Solicitation of Employees.** Commencing immediately, and continuing until a date one (1) year after the date of final completion of the Work, the Integrator and Customer agree not to directly or indirectly employ, solicit for employment, or advise or recommend to any other person that such other person employ or solicit for employment, any person employed by or under contract to the other.

Paragraph 8 is a commonplace term for amendments to contracts. It simply makes clear that the RIDER represents a more recent "mark-up" of the main Contract and therefore takes precedence.

8. **Conflicts in Terms.** To the extent the Contract and this Rider conflict in any respect, this Rider shall govern. All additional, inconsistent or contrary terms or conditions in any other agreement or proposal are hereby rejected by the parties and shall not become a part of the contract between them.

DATED this _____ day of 20___.

CUSTOMER   INTEGRATOR

Signed:_____   Signed: _____

Printed: _____   Printed: _____

Title: _____   Title: _____

Chapter 8

Specifications

Although I have advised numerous automation companies, I don't pretend to be an engineer. In fact, you would probably be within your rights to label me a "liberal arts" control engineering lawyer. What does that mean? It means that even though I may understand much better than your storefront lawyer what you are talking about when you say "system integration," you will never find me writing a line of code. Or let me put it another way: On the odd chance the "lawyer thing" doesn't work out, you can bet your pinstripes you won't find me going to the patent office with the latest breakthrough in visual recognition technology. So why in the name of PLCs do I think I can write specs better than you?

I'll tell you why. Once technical completeness has been achieved, writing good specifications has nothing to do with science and everything to do with logic. It is in this spirit that I offer my Seven Rules for Writing Specifications. These apply equally to the sort of specs written by owners (that are used as the basis for a bid by an integrator) and to the sort of specs written by integrators (that are included within a proposal).

Rule Number One: Push performance responsibility to the other party. If you're the owner, push for the inclusion of more performance specs and fewer design specs. If you're the integrator, do exactly the opposite. A performance spec requires the contractor to install something that does a particular thing. A design spec only requires the contractor to install a thing—without regard for whether the design is satisfactory or not. The more performance specs there are, the greater the amount of open-ended responsibility that is being handed to the contractor. Similarly, the more design specs there are, the more it is simply a matter of building the thing, and the owner or its design consultant can be blamed if the design is somehow inadequate. Of course, whatever protection a control system integrator may get from following a design spec will be limited or nonexistent if the system integrator itself was the author of the design spec or included it in its proposal. (For more on performance specs, see the first "tip" at the end of this chapter.)

Rule Number Two: Whenever performance responsibility cannot be transferred, push design responsibility to the other party. If you're the owner, for every design spec that you must create, keep it general. That way, the integrator will be responsible for making sub-design decisions—and will be on the hook for any mistakes. If you're the integrator, a design spec should be written so that any warranty of design is contingent on the owner or its consultant's providing timely and complete information about desired functionality.

Rule Number Three: If neither of the previous two risk transfer efforts work, try embedding a requirement that the other party must alert you to any problems that it discovers or should have discovered in the specs before or during installation.

Rule Number Four: If you're the owner, include phrases such as "highest quality," "state of the art" and "free from defects." If you're the automation provider, avoid all such benchmarks (especially these) other than the requirements of the specs themselves, but, if pressed, opt for "industry practices or custom" (or something similar) as an alternative standard.

Rule Number Five: Always include a section on commissioning and training in the specs. If you're the owner, this section is all about making sure your people know how the system works—and squeezing every last bit of data about the system from the integrator. If you're the integrator, this is all about ensuring owner cooperation and reaching an endpoint.

Rule Number Six: Define terms when it is in your interest to do so. Keep them undefined (thus defaulting to "industry practices or custom") when it is not.

Rule Number Seven: Write clearly and with good organization no matter what. While it may sometimes be to your advantage to avoid defining a particular term, creating a muddled or cluttered spec really helps no one—except perhaps one constituency…

You guessed it: the lawyers.

TIP: The Power of Performance Specs

Sometimes looking at the legal side of your business is just a matter of figuring out all the things that can go wrong.

Anyone in the automation business who decides to spend an afternoon actually making a list of potential meltdowns usually ends up calling us lawyers—say, by about three o'clock. Or, alternatively, going straight to happy hour.

Of course, that's true in any business. But in the automation business there is a particular scary situation that many others don't face. It has to do with the difference between "make this" and "make it do this."

As Rod Serling used to say at the beginning of the old *Twilight Zone* program, "Imagine if you will...." Now finish that sentence with "a project with problems." The end user is unhappy. The widgets that were supposed to be happily making their way into the world of commerce are being thrown out as unacceptable. The project's SCADA and MFS are not on the same page.

Worst of all, the lawyers have begun to show their ugly faces.

For the companies involved in this mess, how bad can it be? Well, it depends. There is a major division between those fighting over what are called "design specs" and over what lawyers call "performance specs."

Example of a design spec (and I am simplifying here): "Best Integration Solutions, Inc., will integrate the widget production lines with the warehouse automation system in accordance with the detailed design of Premier Consulting Services." Translation: "Make this."

Example of a performance spec: "Best Integration Solutions, Inc. will integrate the widget production lines with the warehouse automation system such that the combined system will operate at a rate of 100 widgets per hour." Translation: "Make it do this."

For integrators, the problem with "make it do this" (performance) specs is that it puts them on the hook for meeting the owner's specific output requirement. If anything goes wrong (e.g., the widgets move more slowly or less efficiently than desired), the integrator is facing a boatload of potential liability. For this reason, integrators would be well advised to avoid performance specs while end users would be well advised to embrace them.

"Make this" (design) specs reverse that situation. The integrator is following the plan or design of a consultant. As long as it installs the exact system that was specified, whether the system works or does not work is someone else's problem. For obvious reasons, therefore, integrators would be well advised to embrace design specs while end users would be well advised to avoid them.

Is there a middle ground? Sure. One way is to pair a performance spec with a hard limit on liability for the integrator. Another approach is to pair a design spec with a set of production-based compensation incentives.

Now, what happens when (as is often the case) there is no detailed contract on these points— but instead an exchange of a one-page proposal and a one-page purchase order? In other words, what happens when "make this" and "make it do this" give way to no clear spec at all? Hate to say this, but it depends. Depends on what was said and what was done. It could also depend on industry custom and the law of the place where the project is being done. In some circumstances, the law could "imply" something called a "warranty of fitness for a particular purpose." What's that? Think of it as an implied "make it do this" spec. Bad for integrators. Good for end users.

TIP: What If the Specs or Plans Are Defective?

What are the legal consequences if an integrator follows a set of specifications or plans provided by the owner—and those specs or plans are defective? A long-standing and widely-accepted legal principle called the *Spearin doctrine* (named after one of the parties in a 1918 court case) may come to the rescue. The Spearin doctrine says that a contract containing specific performance requirements carries with it an implied warranty that if the plans and specs are followed, the result will be adequate. In other words, if defects occur in that situation, the contractor will not be liable. In more recent years, this principle has been extended to products as well—which is to say, a company may not be liable if a product selected by the owner was not suitable for the use that was specified.

TIP: The Pitfalls of Owner-Specified Equipment

You would think it would be pretty simple: The project specifications require installation of an Acme DPFG motor. Your company purchases the product, installs it in accordance with the manufacturer's instructions, and that should be it. The owner ought to be happy—except that's not the case. The DPFG motor is failing at the 16-month point. The motor only comes with a one-year manufacturer's warranty, but you have agreed, elsewhere in the contract documents, to warrant it for two. The owner is looking squarely at you to "make it right."

The general rule, of course, is that an owner warrants its specifications to be adequate to build the project and that if the integrator follows them (blatantly obvious errors excepted), the integrator has done its job regardless of whether the specifications meet the owner's purpose. So, if they contain proprietary specifications calling for installation of a particular Acme product, the integrator ought to be safe regardless of whether the products are suitable for the project. Right? Well ... yes, but not always.

Proprietary specifications can sometimes be inconsistent with other project requirements. The warranty duration mentioned above is a good example. In one of our law firm's recent cases, the owner contended that it had bargained for a three-year warranty and that the amount it had agreed to pay the contractor was intended to reflect that obligation. In the owner's view, it was up to the contractor to obtain a longer warranty from the manufacturer if something less than three years was "standard" (or the contractor should be prepared to pick up the cost of the extra two years of warranty compliance on its own). Never mind that the owner's engineer had been aware that the particular product required by the specs came with only a standard one-year warranty.

Now, it is true as a general proposition (at least where the owner or its consultant contributes specs) that it is the owner's responsibility to determine whether a design is adequate for the purposes intended and/or whether a product specification is appropriate. Nevertheless, many owner/integrator contracts still place verification of design or product appropriateness on the integrator's shoulders, especially when the owner's engineer has not had significant involvement in the project. And in those circumstances, it is not uncommon for the owner to require that a given manufacturer's particular product be used. A lesson from this is that an integrator should never take for granted that the incorporation of a manufacturer's product meshes with all other contract requirements.

Chapter 9

Intellectual Property

Intellectual property (what lawyers call IP) is any idea that can be owned. It can be a device, an invention, a process—even a secret way of doing things. And because ideas are at the heart of all successful businesses, it is no exaggeration to say that of all the legal mistakes an automation-related company can make, those involving intellectual property are near the top of the list.

Here are five easy ways to mismanage intellectual property. Make these mistakes and you stand a decent chance of wrecking your business, missing your opportunity to make millions and wasting a lot of time—and/or paying the tuitions of your lawyer's children.

Mistake No. 1: Failing to recognize and patent your important inventions. Having a thing of value and either failing to recognize it for what it is or even worse, knowing that it is valuable and failing to protect it from others is the most fundamental of IP mistakes. The remedy, in many cases, is a patent.

A patent is a right to an idea that is granted to an inventor by the federal government. It generally lasts 20 years. Interestingly, if you own a patent, you have not gained any rights to do something *yourself* that you could not do before. However, you have gained the right to prevent *others* from doing something—namely, from using, making or selling your invention.

Patents are only granted to individuals, not companies. In practice, however, they are often acquired and owned by companies—or assigned or licensed to the company by the inventor according to a pre-existing agreement. "Assign," by the way, means to give up all rights to the patent. "License" means to give someone else the right to use while retaining ownership of the patent.

But a patent is only one type of protection. The second is contractual (what all the parties involved with the invention agreed).

Mistake No. 2: Creating an invention for a customer without defining everyone's rights ahead of time. While patent protection is wise, there is not always the opportunity (or time) to recognize the idea for what it is and go through the process of filing an application. Nor does filing an application address the prospect of disputed ownership. Let's say you have just given birth to an idea in the midst of a project for a customer. Here are the critical questions:

- Is it the customer's invention now, or is it yours?

- Or does one party own it, with the other receiving a non-exclusive, royalty-free license to use it?

- And if there's a license, can it be transferred to someone else?

- Will a royalty be charged in connection with the license? (For thoughts on pricing, please see the checklist at the conclusion of this chapter.)

All of these questions can be answered within the body of the contract governing the project—for instance, in an IP paragraph like this one:

> **Section 16.01.** Integrator shall retain all right, title and interest in the System and all other intellectual property connected to its Work, including but not limited to all other drawings, specifications and software prepared by Integrator, and all other copyrights, patents and intellectual property rights, except that Integrator shall grant to Owner a nonexclusive license to install and use the System.

What happens if the customer then ignores the contract and sells the idea to a third party? Answer: the inventor has recourse against the customer—meaning a lawsuit may be successfully pursued (with the level of success dependent on other paragraphs of the contract). If you are wondering about the inventor's rights against the third party—now, that is a different sort of problem. Certainly the absence of a contract between the inventor and the third party creates a major barrier to taking action. What would bring down that barrier? You guessed it—a patent.

Mistake No. 3: Infringing on someone else's patent. If you want to see your company suffer huge losses in a single year, or even worse, self-destruct, find a way to infringe on someone else's patent. Lawsuits for patent infringement are guaranteed to cost a company hundreds of thousands of dollars or even much more to defend, and may cost additional millions if infringement is found to have occurred. Thus, before you invest too much money in developing a new product or idea, investigate whether the idea infringes upon someone else's patent.

Now maybe this does not apply to you because you are a risk taker. Maybe you've got the notion that you will go ahead with the development of your new idea without spending money on an investigation, and then later claim, if sued, that you had no notice or knowledge of the infringement. This is more than a risky course of action, however—it is downright reckless. As with many areas of the law, ignorance is no excuse, and the patent infringer is unlikely to receive any lenience. On the contrary, the infringer's lack of investigation will likely be labeled an intentional infringement in a court of law, and may be punishable with treble damages, which is lawyer-speak for a tripling of the loss the patent holder actually had.

Does your company need to hire a lawyer to protect itself? Not necessarily. It is possible to get a flavor for what patents have been issued in a particular area by searching on the web site of the U.S. Patent and Trademark Office (www.uspto.gov). There, you can search for issued U.S. patents and patent applications by keywords that describe the new product or idea. A slightly more sophisticated search may be performed by searching in specific classifications or sub-classifications for inventions. Performing such a search is relatively easy and free of charge and can provide the searcher with a general idea about whether an invention has already been patented by another party.

The best and surest protection from patent infringement is consulting a patent attorney, who can have a more extensive prior-art search performed and who can provide a formal opinion as to whether the new product or idea may infringe upon an existing U.S. patent. Such an attorney typically begins this work by forwarding a description of the new product to a professional patent searcher who is geographically located near the records of the U.S. Patent Office in the Washington, D.C. area. The searcher will consult with examiners from the Patent Office to determine the most fruitful classifications in which the search should be conducted. Further, the searcher will peruse the archived paper records of the Patent Office in search of patents that may not be available in electronic format on the Patent Office's web site.

Upon receiving the search report, the patent attorney will closely examine the claims of each patent and compare the claims to the new invention. If infringement problems are found, the patent attorney may counsel the company as to how it may modify its invention to "design around" the problematic patent and thus avoid any claims of infringement. If a "design around" is not possible, however, the company has the opportunity to negotiate a license with the patent holder—or abandon the infringing idea before too much money has been invested.

Because reasonable minds may differ as to whether two technologies are similar, a patent attorney's non-infringement opinion does not guarantee that infringement will not later be found in a court of law. However, an additional advantage of getting a non-infringement opinion is that it is a good way to demonstrate good faith and therefore avoid the worst result (treble damages) if patent infringement is later found.

Mistake No. 4: Licensing your technology while neglecting or ignoring background patents. This is a subspecies of the infringement problem. A background patent is a patent that is a building block or foundation of a second patented invention. In other words, Patent A (someone else's) must be used before Patent B (yours) can be used. This is a problem of great relevance to the automation industry because control system software is often built upon the functionality of pre-existing software owned by someone else.

The essential questions here are the following:

- Does your underlying software license authorize you to sublicense?
- If so, at what cost?
- For how long?
- With what obligations?

Or putting the problem a bit differently as a customer-driven proposition: What will it take to assure the licensee of your invention that some third party will not surface later and try to spoil things?

Mistake No. 5: Neglecting all of the many other important details. Although an automation company may have protected its valuable inventions by patenting them (and protecting its nascent ideas via contracts and licensing), it may nevertheless stumble if it neglects the details.

One important set of details can easily be overlooked if an automation company makes the false assumption that the world is an unchanging place. Change, of course, happens. The company that was granted the original license may have been acquired or sold, or otherwise changed hands. Its business, markets or customers may have shifted. The licensee may have tinkered with the invention, changing its nature or transforming it altogether. Or the machines making use of software may have been sold. When even a little attention is paid to some of the issues that *can* arise over time, it is not difficult to create a list:

- Does the license transfer with the company or individual machine?
- Is the licensee required to give notice of any transfer?
- Does the inventor have the right to approve or reject any transfer?
- Must the inventor give its OK to each sub-licensee?
- Is the sub-licensee required to accept the license terms?
- May the licensee adapt or improve upon the original invention?
- Must the licensee disclose changes to the inventor?
- Does the inventor have the right to incorporate changes in future versions of the invention? For a fee? At no cost?
- Can the inventor share improvements with other licensees?
- Does each licensee have access to source code or other secret components?
- Is the scope of the license limited to a particular system or machine?
- May duplicate hardware be assembled and the invention used at additional machines at the same site?
- Is the license unlimited?
- Will the inventing company be granted a security interest in a machine for nonpayment of royalties?
- Does the software disable itself annually, or quarterly?
- Is there a nondisclosure agreement for the operating manuals you provide?

Most, if not all, of the glitches that you can imagine in the *absence* of giving attention to these details can be eliminated by some investment in better defining the front-end agreement. Ignoring these potential problems, on the other hand, based on a belief either that the customer-contractor relationship is strong (and the details can be worked out later) or that the likelihood of such conflicts is remote is a much riskier strategy.

CHECKLIST: Six Questions to Ask when Setting Automation License Royalties

- ✔ What is the useful life of the hardware-software combination?
- ✔ What is the cost of developing the hardware-software combination?
- ✔ How should the royalties be structured?
    - A one-time lump-sum charge that covers both the labor and materials component and the IP component
    - A one-time payment for the labor and materials component and an ongoing royalty for the IP component
    - Net sales value of goods manufactured and sold using the machine and/or software
    - Volume of goods
    - Other measures (e.g., weight)
- ✔ Should there be an annual minimum royalty?
- ✔ Should maintenance or service be rolled into the license or be addressed in a separate transaction?
- ✔ What is the duration of any royalties?
    - Should they be limited to the life of the patent?
    - Should they be for a specific term of years?
    - Should they be functionally-determined (e.g., for as long as the machine and/or software are being used)?

Chapter 10

Automation Standards

A standard, according to one popular dictionary, is "something established by authority, custom, or general consent as a model or example." By that measure, the automation industry is rife with standards.

You might not be surprised to learn that a number of these have specific legal consequences.

I recently took an unscientific poll of automation professionals via various social media groups to ask a simple question: What are the automation standards that you encounter most?

There were the inevitable non-answer answers: "I think you need to refine the question." "Any answer you get will be meaningless." And my favorite: "I didn't realize that automation was being implemented by lawyers. But I suppose I should have."

But then we got into the meat of it:

- "From a formal standards perspective, I rely on ISA standards where they exist. From there, I fan out to IEC and other accredited standards."

- "Definitely ISA-88 and its international shadow IEC 61512."

- "Mainly ISA and IEC. [In] a minor role: NEC, IEEE, PIP."

- "IEC 61131 PLC standards and IEC 60417 graphical symbols for use on equipment."

- "If you're looking for quality of service issues, I would say the CSIA Best Practices & Benchmarks is a great place to start."

- "When it comes to technical work, I would have to say NFPA 70, 70e and 79. These relate to the National Electric Code and are used by most all enforcement agencies, including OSHA."

- "Almost always: UL 508A, NFPA 79, state/local electrical codes."
- "I think that ISA-95 has much broader applicability to the manufacturing community."

I used some word frequency mapping software to put in graphic form (by size) the terms that appeared in the answers, as shown in Figure 5.

Figure 5—Frequency of Words (by Size) in Responses to Informal Automation Standards Survey.

There were some recurring acronyms, of course: IEC, ISA, NFPA, CSIA. That's when it hit me: You need look no further than the standards promulgated by each of these four organizations to see some important distinctions in the way the law treats them—not to mention confirmation of the premise that not all automation standards are created equal.

In this sense the individual who wrote "Any answer you will get will be meaningless" had it right. You cannot really assess a standard until you have determined its category. Here are the four categories as I see them:

Spec Standards

IEC 61131-3 strikes me as a pretty good example of a spec standard. This standard, from the International Electrotechnical Commission is said to be the first vendor-independent programming language for PLCs. But, putting that claim aside, there is nothing that requires automation companies to give it any heed if it is not specified. It has not become, at least to my knowledge, a law or regulation in any state. For that reason, I call it a spec standard—the type of standard that is only effective in your project if required by the contract documents. (End-user-created standards are perhaps the most common type of spec standard.)

Spec standards are always enforceable by a customer if they appear in a contract—unless waived or in conflict with other parts of the contract.

Industry Standards

The ANSI/ISA-95 standard fits the definition of what I would describe as an industry standard—which means, in my lawyer mind at least, that it potentially has application to your project even if it is not specified in the contract. That is because this pervasive standard, from the International Society of Automation (ISA), defines best practices for integrating enterprise and control systems. It also is not a law or regulation anywhere (as far as I can tell), but it is nevertheless influential. So when I say that ANSI/ISA-95 potentially has application, what I am really saying is that if your own method of integrating these systems has problems, someone may hang this standard around your neck. Thus, something is an industry standard if it has the potential to be used against you even if it is not in your contract.

However, industry standards are typically not enforceable if they are not mentioned in the contract and the contract has a "lid" on it (what lawyers call "integrated"). This simply means that there is a provision in the contract that says something to the effect that the parties have no other agreement than this one and/or that the contract documents supersede all prior discussions and agreements of any kind.

Code Standards

NFPA 79 is an example of a code standard. This means that not only is it a standard that has attained pervasive influence (such as an industry standard), it also has become—quite literally—the law. This particular standard, from the National Fire Protection Association, governs electrical wiring in industrial machinery. But you don't need to contact the NFPA to find it. It's right there in the code books. Although the versions adopted by state legislatures and localities are typically a few years behind the most current version, this is only a minor problem. The bottom line is that code standards do not need to be in the contract to have legal consequences. If you ignore NFPA 79, there is a good chance you have broken the law.

Code standards are always enforceable by the applicable governing authority and do not need to appear in the contract to be binding. In many cases they are enforceable by the customer as well. Code standards are difficult, if not impossible, to waive.

Aspirational Standards

At the other end of the spectrum, I see the *Best Practices and Benchmarks* used by the Control System Integrators Association (CSIA) as the paradigm of an aspirational standard. CSIA standards, which cover a range of activities from business basics to project competency, in my view set apart CSIA-certified integrators as being the best in the industry. But they are not the sort of standards that are specified for a project—and end users rarely see them. This industry-internal quality makes them, in my view, purely aspirational (if applied correctly).

Moving from the most binding to the least binding category of standards, start with code standards (always binding) to spec standards (binding if specified), then move through industry standards (persuasively binding) to aspirational standards (not intended to be binding). This spectrum is shown in Figure 6.

Chapter 10 Automation Standards

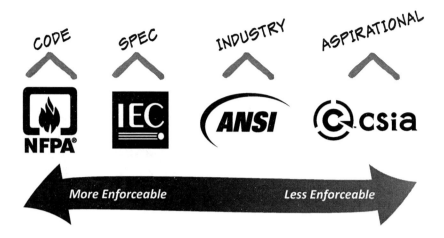

Figure 6—The Spectrum of Automation Standards.

Only by recognizing what type of standard you are reviewing can you understand its legal impact. Keep in mind there are numerous standards that can cross over between categories—depending on the contract in question or the governing authority—such as those for lean manufacturing, the ISO 9000 series of standards and LEED. See the table in Figure 7 for my (admittedly subjective) slotting of these and other examples of standards.

NAME	SUBJECT MATTER	TYPE OF STANDARD
Lean	Preserving value with less work	Spec, industry, aspirational
ISO 9000	Quality management systems	Industry, aspirational
LEED	Energy sustainability	Code, spec, aspirational
OSHA 1910	Safety on worksite	Code
IEC 61508	Functional safety of programmable electronic equipment (SIL levels)	Code (EU); industry
HACCP	Food safety	Code, industry
UL 508 (transitioned to UL 60947 series)	Industrial control panel standards	Code-related, industry
ATEX directive	Equipment in a potentially-explosive environment	Code (EU)
IEEE 802	LAN standards	Industry (but code-like)
PIP	Process industry practices	Aspirational, industry, spec

Figure 7—Categorization of Some Automation Standards.

Classifying any potentially-applicable standard is just one step in what ideally is a multi-part analytical process. In my view, the most effective approach for project management is the following sequence:

- **Identify the contract.**
    - Locate all of the documents that comprise the contract.
    - Determine whether the contract has a "lid" on it—what lawyers call "integrated." (In other words, does it have a clause that makes clear that these particular contract documents are the only ones that apply?)
- **Identify any standards that are indisputably identified in the contract or arguably applicable to the type of work being done.**
    - The *indisputably* incorporated standards are code standards and spec standards.
    - The *arguably* incorporated standards are industry standards and aspirational standards.
    - **(Standards that are only arguably applicable—industry or aspirational—have a greater likelihood of applying if the contract is not integrated).**
    - Ask yourself whether there are any conflicts between standards that ought to be resolved.
- **Modify the contract if it is advantageous to do so.**
    - Remove any inapplicable or problematic references to standards.
    - If part of a standard is relevant and part is not, remove any reference to the part that is not relevant.
    - Add conflict resolvers (a default order-of-priority clause that decides which standard prevails in the event of conflicts).

In my view, it is a bad practice for either end user or vendor to include in the contract a long list of standards that may or may not be applicable to the project in question. For the vendor, doing this can create an inhospitable landscape of obligations that makes spec compliance nearly impossible. For the end user, it increases the possibility of conflicts that can create confusion at a minimum—and in the worst case can neuter those standards that are the most important (even code standards, if the relevance of the code standard to the application is less than clear).

Chapter 11

Professional Licensing

Once while speaking at an automation conference, I took a moment away from my talk to poll the audience. "Stand up," I said, "if your company works on control systems." Most everyone stood up. "Remain standing if your company employs at least one person with an engineering degree." A number of the executives sat down. "Remain standing if your company employs at least one person who is a licensed professional engineer." Half the room sat down. Finally I said, "Remain standing if your company employs at least one person who has passed the Control Systems Engineer (CSE) exam." Only a handful were left standing.

What does this mean? In the event that you are one of the few who did not already know the answer, the connection between professional licensing and the automation industry is tenuous at best. Is this a problem? My answer is "No," but there is an asterisk.

Here's why I say "No." It mainly has to do with a creation in North America called the "industrial exemption." This exemption, which means different things depending on the state or province you operate in, more or less amounts to a decree to government agencies to keep their noses out of assembly lines and process facilities ("Don't stop industry from moving forward"). In other words, long after government regulators decided that everything from embalmers to lawyers to hairdressers to furniture retailers (someone explain this one to me) needed to be licensed, industrial technicians were left alone. So there is something of a tradition of industrial engineering being unlicensed.

Two other influences are also helping to preserve the unlicensed environment, in my view: safety in numbers ("Wait—are you telling me you are proposing to put all of these unlicensed automation companies out of business?"), and even more significantly, the inability of government to keep up with technology (e.g., software engineers are largely unlicensed—and software is obviously a huge part of control engineering).

But the "asterisk" part of my answer cannot be ignored either. There are four reasons why automation companies should at least have professional licensing on their radar screens:

First, there are sporadic efforts to eliminate the industrial exemption (and, to be sure, regulation of this market segment could be just one engineering catastrophe away).

Second, the current definition in most states of what constitutes the practice of "professional engineering" (i.e., PEs) seems to fit what persons in the automation industry do—even if the extension of the PE label to the industry is not widespread.

Third, at least one state, South Carolina, has clamped down on unlicensed engineers who are engaged in the profession of control engineering. In a June 2007 ruling, the South Carolina licensing board made it crystal clear that "systems integrators do need licensed engineers, qualified to provide control systems design, on their staff."

Fourth, the consequences of noncompliance with a professional licensing requirement—the potential shutting down of a business—even if remote, may not be worth the risk.

Interestingly enough, it is not the risk of a state agency imposing a fine or shutting down a business that should be the chief concern. Instead, it is the prospect that a customer could use the absence of a license as an excuse not to make payment. Here is the scenario: Customer hires integrator to design and install a control system. The system is completed, but a dispute arises from some problematic feature of the system; as a result of this, the integrator is not paid. In the lawsuit filed by the integrator, the customer defends its position by asserting, among other things, that the integrator is not entitled to be paid because its lead professional was not a licensed engineer.

Believe it or not, there are court decisions in more than one state that back up that argument—with rulings that a company without a licensed engineer could not enforce its contract in court.

CHECKLIST: Nine Realities of Professional Licensing for Automation Companies

- ✔ Professional licensing is permission from the government to do business in a specialized area (usually an area where there is a significant downside to unqualified work being performed).

- ✔ State licensing requirements are in place for numerous professions, from hairdresser to embalmer to furniture retailer. For automation professionals, the key type of licensing is that of the Professional Engineer or PE.

- ✔ In most states, the definition of "professional engineer" would appear to fit what control system engineers do.

- ✔ Even so, only a few state or local agencies mention control system engineering in their statutes or regulations.

- ✔ South Carolina is the only state (to my knowledge) that requires control system integrators to employ licensed engineers.

- ✔ The ISA's Control Systems Engineer (CSE) examination (a type of PE exam) is widely available for interested applicants across the country.

- ✔ There are two risks of not being licensed. One of them is being fined by a professional licensing agency.

- ✔ The other risk is not being paid. Some states permit a customer to refuse payment if engineering is done without a professional engineering license.

- ✔ A long-standing "industrial exemption" in North America may provide protection in some circumstances to unlicensed automation professionals.

TIP: How Licensing Works in the U.S.

If a professional engineer (PE) license is required or desired for work in the U.S., there is a well-worn process. However, with every U.S. state having its own rules for the licensing of engineers, navigation of that process sometimes can be challenging.

The place to start is the licensing board for the state where most work is done. Most states require graduation with a four-year degree from an accredited engineering school as a threshold matter. There are then two levels of testing (customized by each state) that are drawn from a standardized set of tests developed (and graded) by a national organization, the National Council of Examiners for Engineering and Surveying (NCEES).

The first test is the "Fundamentals of Engineering" (FE) exam. Passing of this exam typically confers either the title of "engineer-in-training" or "engineering intern." If the state is a "discipline state" (not all are), the FE exam might include testing in specialized areas (such as civil or electrical—or, rarely, controls). After engineering experience is gained (typically after four years), the second exam, called the "Principles and Practice of Engineering" (PE) exam, can be taken. The second level of exam is nearly always focused upon specialized knowledge.

This is just a general outline and there can be ways of short circuiting the process. In some states it may be possible to bypass one or more of the prerequisites based on experience or if a license is held from another state.

TIP: How Licensing Works Internationally

Licensing of engineers in countries other than the U.S. vary drastically. For instance, although licensing in Canada is similar to the U.S., the United Kingdom does not regulate engineering at all (although a chartered certification is available). This is also the model across much of Europe as well, with the professional title of engineer being more a matter of certification or qualification that can be used as leverage in business as opposed to being a legal requirement imposed by government authorities.

Chapter 12

"Green" Considerations*

Many businesses today are responding in one way or another to the increasing social, political and economic pressure to "go green" by implementing processes and systems that reduce energy consumption, conserve water, and decrease any adverse environmental impacts caused by their methods. The automation industry is no different—after all, automation is all about efficiency, and why shouldn't already process-efficient systems be environmentally efficient too?

Integrators and OEMs can feel the pressure to go green from a number of different sources, but the most likely one is an end user who now wants an environmentally efficient control system in addition to a system that efficiently completes the desired process.

Some of the direct monetary reasons that a customer will press for green systems include cost savings from reduced energy consumption and potential tax incentives offered at various levels of government for achieving a predetermined level of greenness. Indirectly, social and political pressures also lead customers to believe that there is a fair amount of goodwill to be earned through advertising that their company is green, and as a result future economic gain is a likely secondary benefit (given a choice, and all things being equal, consumers will pick the green company because it makes them feel good). Recognizing these potential benefits is important because failing to achieve the desired—and possibly contracted-for—level of greenness can have a real cost (see below).

But first, how is a manufacturing company's greenness measured? From a scientific perspective, you might compare the amount of water, energy, and fuel used, the air quality, and any pollution created both before and after implementing green measures to determine whether the same task is being com-

---

* The author thanks his colleague, Shawn Doorhy, for his contribution to this chapter.

pleted in a more environmentally efficient manner after modification. But claiming greenness with that level of detail is too complicated, too difficult to advertise or standardize, and there are any number of variables that could skew the results. How, for example, could anyone verify a company's claim that they have reduced the amount of electricity needed to operate their control system by 30%?

To standardize this type of analysis and provide some credibility to claims of being green, several third-party organizations and programs have entered the marketplace and offer to certify a customer's building or processes as being environmentally friendly at defined levels. Customers often rely on these third-party certifications as a standard by which to advertise greenness and governments use them to determine whether the qualifications for a particular tax benefit are met. The most prominent examples of these organizations and programs include ENERGY STAR, which is a government program that rates products and buildings, the United States Green Building Counsel (the USGBC), which offers its LEED (Leadership in Energy and Environmental Design) certification program for buildings, and the Green Building Initiative (GBI), which offers its Green Globes certification program, also rating buildings.

Figure 8—Three Types of "Green" Certification.

ENERGY STAR

Created by the Environmental Protection Agency and Department of Energy in 1992, ENERGY STAR was initially developed as a labeling device for products in order to promote energy efficiency. Chances are you peeled the now famous ENERGY STAR sticker off your first CRT computer monitor, tossed the yellow ENERGY STAR information card that was taped to the side of your new refrigerator, and overlooked the star logo on the box of your new flat screen television—which is to say, ENERGY STAR is everywhere. In fact, much of the hardware used in control systems today probably carries this designation for some or all of its components. The program recently added designations for homes and buildings, but is different from the LEED and Green Globes systems discussed below because it only awards the ENERGY STAR label. For products, the EPA chooses which ones obtain the label based on a set of guiding principles relating to energy efficiency. For buildings, a particular building must score a 75 or higher on a 100 point scale that compares energy consumption for all similar buildings nationwide in order to use the label. The use of ENERGY STAR labeled products and materials in buildings helps earn more points in all three systems.

LEED

The USGBC's LEED rating system, which is administered through its affiliate, the Green Building Certification Institute (GBCI), rates a building's design, construction and performance with a point system, awarding points for environmental factors falling into in six different categories: Sustainable Sites, Water Efficiency, Energy and Atmosphere, Materials and Resources, Indoor Environmental Quality and Innovation in Design. The levels of certification a building can achieve are: LEED Certified (requiring 40-49 points), LEED Silver (50-59 points), LEED Gold (60-79 points), and LEED Platinum (80 points or above). Individuals can also become LEED accredited by taking a series of exams on the rating system and how it works.

Don't be fooled by the LEED rating system's focus on certifying only buildings and not products. LEED also examines the processes going on inside a building. That means the environmental efficiency of the control system installed has a direct relationship with the environmental efficiency of the building in which it is contained, thus impacting the building's LEED rating. A building can be certified as LEED under several different programs tailored to the type of building and the manner in which it will be used. For automation purposes, the LEED for New Construction and LEED for Existing Buildings: Operations & Maintenance programs are likely the most applicable.

Green Globes

Similar to and in direct competition with the LEED system, Green Globes rates the energy efficiency of buildings on a point system, but awards Globes on a percentage basis of the points earned out of an available 1000 points. One Globe requires 35-54%, two Globes requires 55-69%, three Globes 70-84%, and four Globes 85-100%. With Canadian roots, Green Globes has more use internationally but is gaining a foothold in the United States as its standards are increasingly being used by the American National Standards Institute. One advantage of Green Globes over LEED is that by using a percentage based rating, the system is more flexible for projects that have an increased focus on one environmentally-related aspect but for which others are not applicable.

So what legal concerns arise from the green trend? Consider this example: End User Distribution, Inc. issues a purchase order to Best Systems Integrator, Inc. for an automated conveyor system that will sort and transport pre-labeled packages from a storage area in an existing warehouse to an adjacent loading dock and then onto trucks for distribution. The P.O. calls for BSI to provide new system software and computer hardware and to integrate it with an existing conveyor system. In addition, the P.O. includes a performance specification: it must use 20% less energy and emit 30% less heat than the prior system. In a separate conversation between EUD's chief operating officer and BSI's project manager, the COO reveals that EUD is pursuing LEED Silver certification for the building which, when obtained, will earn EUD a $2.3 million tax break with the local government and will position EUD favorably for a lucrative package distribution contract that requires use of a LEED Silver warehouse.

At the end of the project BSI's system flawlessly sorts and moves the packages from point A to point B, and does it at *twice* the rate of EUD's prior system. The PM (project manager) thinks this is phenomenal. However, despite hundreds of hours of work BSI is only able to get the system to use 15% less energy and emit no less heat, which the PM thinks won't really concern EUD given the doubled process speed.

Rather than being elated about the improved speed, EUD slaps BSI with a $5 million breach of contract lawsuit. How is this possible, you ask? Well, EUD claims, because BSI was unable to meet the required reduced energy and heat specifications, EUD's building failed to qualify for the LEED Silver certification.

That cost EUD the $2.3 million tax credit, another $1.7 million in expected profits from the lost distribution contract, and another $1 million in expected energy savings over the expected life span of the entire system. Suddenly greenness has a real cost to the integrator and end user, and it adds a new level of thinking for the lawyers involved—for example, BSI's LEED accredited attorney might respond to the lawsuit by claiming that EUD failed to mitigate its damages because it could have purchased renewable energy credits for $500,000, which would have earned EUD enough points in a different category under the LEED system to qualify for LEED Silver.

Think back to the chapter on the Dirty Dozen contract clauses. If consequential damages are not somehow waived or curtailed in the contract, the customer's damages for inadequate performance can easily skyrocket based on failed green objectives alone. Also, keep in mind that LEED, Green Globes, ENERGY STAR, and other green certifications are administered by third parties over whom neither the integrator nor the end user has any control. Rating systems can change (and perhaps become more stringent) in the middle of a project. As a result, no automation contract (or any contract for that matter) should promise or require that such certifications will be earned as a condition of one party's performance.

At the completion of the project, any improved environmental efficiency will almost certainly need to be maintained continuously. A building with an integrated automation system might qualify and obtain LEED Silver, two Globes, or some other rating at the outset, but it will also be expected to continue to perform in such an environmentally friendly fashion (and might be regularly tested on that basis by the third-party organizations described above). This adds another dimension to warranty provisions: not only does the system need to continue to do what it is supposed to do but the same environmentally friendly metrics will also have to be maintained. Failure to do so can produce liability. If, over time, the installed system uses an increasing amount of energy, the green certification might be lost, and the end user might suffer monetary losses as a result: it might have to pay back a portion of any tax credits received, or it may have increased energy costs. Accordingly, if green considerations are part of the project, it is important to ensure that all terms of the contract (not just the scope of work) appropriately account for those considerations.

> ## CHECKLIST: Six Reasons Automation Companies Need to Speak Green
>
> ✔ **Inclusion.** Green building accounts for more than one-third of all non-residential design and construction and will grow to more than one-half of all construction by 2014.
>
> ✔ **Growth.** Green building has been one of the few bright spots in an otherwise dreary construction market.
>
> ✔ **Efficiency.** For example, converged and standardized cabling requires less material and fewer pathways, thus saving energy.
>
> ✔ **Economics.** Another example: HMI-displayed energy efficiency readouts may assist the owner in earning LEED-based tax credits.
>
> ✔ **Liability.** If a green measure or piece of equipment under performs, thus endangering green certification, the automation company may be liable.
>
> ✔ **Responsibility.** Removing waste of motion, time, energy and resources from every step in a process helps the environment.

Chapter 13

Changes and Other Mid-Project Communications

At a conference for automation executives in Santa Fe I spoke on a serious topic, but one with a tongue-in-cheek title. The title was "How to Write Letters and E-Mails That Don't Drive Your Lawyer Crazy." Following that talk, I probably received more requests for my slides than after any other automation law talk I have given before or since.

I should not have been surprised. Integrators, OEMs and process end users—as technically-minded as they all are—typically don't think much about the legal effect of their words until they find themselves on the business end of, say, a subpoena. Now, don't get me wrong—I'm not grousing. As a lawyer, being the clean-up guy on a project that has gone awry keeps my kids in college. But, like all lawyers, I like to be on the winning side—and what my automation clients put in their written communications can greatly affect whether a win (however you define that term) is even possible.

But win *what* exactly? Well, in most cases, I am talking about winning the most important thing that can be discussed during a project—changes. Changes in schedule. Changes in quality. Changes in scope.

Here, in a nutshell, is the issue: Although the general public's conception of the legal process mainly consists of people testifying on the witness stand, the reality is much more tedious than that. There are literally a hundred days spent debating the statements written on paper—or electronically—for every *Law and Order* cross-examination moment. Why is that? It's because what people say in correspondence and memos at the time the events are happening—not after the lawyers get involved—has much greater credibility.

Which brings us to e-mails. E-mails, in particular, are popular targets of lawyers because they tend to be more informal—and less well-thought-out—than paper correspondence. (It's even gotten to the point where courts have created new rules governing the disclosure of e-mails to an adversary in a lawsuit, and many big law firms—mine included—have set up groups of lawyers with special expertise in electronic discovery.)

So what's an automation company to do? I would offer the following suggestions:

First, put a clause in your contracts or specs that makes clear that only certain persons have the authority to speak on behalf of the company.

Second, establish an internal policy for sending out important (or semi-important) project notices that makes it clear that not only do the notices have to come from the project manager, but they also have to be approved (for example, by the project manager's boss).

Third, for larger or more important projects, establish a "run it by the lawyer" policy (which means that you first write the draft e-mail, run it by the company lawyer, then send it out after it has been properly sanitized. This really goes for anything put in writing—e-mails just tending to be the least reflected upon.).

How should e-mails and other written matter be sanitized and/or strengthened? There are a few simple rules:

- They should be consistent with the "story" that your company would want to tell about the project if a problem were to erupt and then need to be resolved by a third party.

- Ideally, they should be easy to understand (i.e., non-technical to the extent possible) for the same reason.

- Where it makes sense to do so, relate your messages regarding project activity to a contract term.

- Rarely paraphrase contract terms. Quote the key parts word for word.

- Avoid making statements of fact that an opponent may use against you if there is any chance you do not yet know or fully understand all the facts.

- Get into the practice of using the phrase "among other things" when listing items, problems, tasks etc. That simple device may save you later if you inadvertently leave out an important concern.

- Respond to all accusations. Failing to respond to an adversary's accusations during a project is worse than responding ineffectively or badly. Your silence can be interpreted as agreement with the facts stated in the adversary's e-mail.

- Reserve your rights. Use a phrase that makes it clear that you are not giving up any rights. It need not sound like it was written by a lawyer. For instance, when trying to work out a dispute, you might say, "In proposing this compromise, we are reserving all rights under our contract if we do not settle our differences."

Now, despite the dangers of an inopportune statement in a hastily transmitted e-mail, the conclusion that you reach from this should not be that you might avoid all problems by shutting down the practice of sending e-mails altogether. That would be a mistake—and would indeed create your lawyer's worst nightmare: a legal confrontation in which the only contemporaneous account of what happened was written by the enemy!

TIP: Paying Attention to Fine Print at the End of a Project

One area where it pays to be watchful *at the end of a project* is fine print that may seem innocuous or even favorable on the surface, but in reality, may be little more than an underhanded attempt to impose an undesirable settlement. Three examples:

- An owner or general contractor tenders a check to your company for *part* of the balance that you believe is owed on a job, and the accompanying letter or notation on the check suggests that it is "final payment." A subcontractor in Virginia fell for that trick and found (after years of litigation) that its cashing of the check was what lawyers call an "accord and satisfaction." Translation: the cashing of the check showed the subcontractor's agreement to the lesser amount. (By the way, that trick did not work in a similar Virginia case where the "final payment" words appeared on the back of the check and there was no evidence that anyone working for the subcontractor ever actually saw the words.)

- A disputed change order is processed after long delay, and it's time for your company's signature on the dotted line. The only problem is that the fine print provides that acceptance of the change order represents a waiver of any and all claims through that date, including claims for delays or disruption. Depending on the circumstances, it may not be wise to sign!

- Even suppliers can get in on the act. Did you know that a supplier's repeated transmittal of an incorrect invoice for payment (and your receipt of that invoice without objection) can, over time, constitute an "account stated?" What's an "account stated"? Basically, it's an implication drawn by the law that your repeated silence in the face of the invoice is an admission that the invoice is correct. In other words, even *doing nothing* can be costly!

The bottom line: if you are not sure about the legal impact, it may be worth a call to your attorney.

TIP: Managing Changes on Automation Projects

It has been said that construction contracts are more challenging than other types of contracts because, unlike the other types, construction contracts are being written (or re-written) at the same time as they are being performed. While a bit of an exaggeration, the underlying point is a worthy one. It simply acknowledges that modifications of contract terms (most typically in the form of agreed changes or as a result of unforeseen events) are not the exception to the rule, but *are* the rule. Indeed, it would be a most unusual construction project (perhaps there are none) in which the original specifications and terms and conditions remain unchanged through final completion.

Take that reality and multiply it times ten and you will have a sense for the multiplicity of changes that automation companies typically must confront. Automation projects, by their nature, literally are constantly changing. Accordingly, for maximum success in managing automation risk one must find a way to manage this anticipated, but frustrating state of affairs. The flood of changes can be managed in at least two ways.

The first way is to control the definition of change. Because the final scope of the software component of an automation project is, in particular, an inherently moving target, it is in the interest of owners and end users to limit the definition of change to a certain threshold or criteria just as it is in the interest of control system integrators and engineering firms to avoid any such limitation while carving out a point where excessive changes are deemed a breach of the contract (in traditional construction parlance, this is known as a "cardinal change").

The second way is to control how adjustments due to change are processed. Owners and end users will want to impose strict notice criteria and tight timeframes for seeking adjustments. Integrators and engineering firms, on the other hand, tend to benefit from relaxed or nonexistent notice provisions and deadlines.

Because profound changes are so predictable on automation projects, the importance of addressing changes on the front end cannot be overemphasized.

> ### TIP: Do Too Many Changes Open Up an Entire Automation Project to Renegotiation?
>
> Can there be too many change orders, or change orders that are too big, on an automation project? The idea of the "cardinal change" doctrine is that there is a breaking point on any job, and when that happens, the original deal is off and a new deal is on. But is the doctrine used very often? Perhaps not. That is because in many, if not most, projects, such problems are handled within the ordinary change order framework.
>
> So, assuming that the jurisdiction in which your company is working recognizes this doctrine, when would it ever need to be used? Maybe in the situation in which the change, or cluster of changes, are so outside the scope of the agreed upon work that the entire project can be said to have been completely transformed. In that case, the integrator is entitled to abandon the contract and is no longer obligated to perform.
>
> Of course, it takes a lot of grit to take that course. Because there is no bright-line test to distinguish a cardinal change from those within the original scope of the contract (and within the owner's right to order), and because abandonment of a project can have serious, and risky, consequences, an integrator typically will not be willing to risk abandonment of a project in the face of even unfair or burdensome changes. However, the cardinal change doctrine can still be of use to the integrator. If it elects to remain and perform the changed work under protest, it may be entitled to recover the "fair and reasonable value" of the work plus the cost of the resulting extension of time regardless—notwithstanding contract language to the contrary. This can be particularly useful when an integrator is faced with such things as a contractual "no damage for delay" clause or strict contractual notice of claim provisions that have not been met.

Chapter 14

Dispute Resolution

Although for many years I have owed my livelihood to the American civil system for resolving disputes, I would not count myself as one of its most ardent admirers. I admit that I have not made a close study of any other country's legal system; for all I know there is no better one. Even so, I have become increasingly troubled by the expensive (and frankly, inefficient) gauntlet that participants must travel to get from the beginning of a dispute to its resolution.

The importance of understanding this reality for anyone contemplating suing or arbitrating should not be underestimated. The litigation process nearly always costs more than it should and takes far too long. Sometimes the expense is worth it, but all too often it is not. All too often litigation becomes little more than a game of "chicken," in which neither participant is willing to yield to the other—but the worst possible outcome occurs when no one gives in. All of this has led to my own personal conclusion that litigation should be avoided if at all possible.

There are always exceptions, of course—and I can think of at least four.

The first is when the stakes are so high that they dwarf the costs. For instance, when the *Deepwater Horizon* disaster occurred, there was very little room for weighing the costs versus the benefits of litigation. It was simply inevitable that there would be lawsuits. (Heavy construction is also in this category—the numbers are so big that the lawyer fees are relatively minor by comparison.)

The second is when the opponent is irrational or misguided—or just hell bent for litigation. There is little choice involved in going or not going to court when the other party will not take "no" for an answer.

The third is when the litigation is part of a larger picture that cannot be understood by looking only at the dollars and cents at issue in the matter at hand.

For instance, one legal result may affect another, or it may be important to send a message or draw a line.

The fourth is when there is a source for paying (or reimbursing) the legal fees if you win. This can be as a result of local laws (such as mechanic's lien laws) or contract provisions that shift the responsibility of paying to the loser of the lawsuit. Or, even better, when the lawyer is willing to take the case on a contingency basis (as TV personal injury lawyers put it, "I get paid when you get paid").

As for all other lawsuits—the vast majority, I would argue—the best advice is to forget it. Any principle that is so important that it justifies a lawsuit tends to crumble under the weight of months (or years) of legal bills.

If one is trying to specifically identify the places where the civil litigation process wastes time and money, one could justifiably lay the blame at two doors. First is the very essence of the advocacy system itself. Second is the "American Rule" for reimbursement of legal costs. While these are not the only sources of unwarranted expense and inefficiency, they are a good start.

Advocacy System

Throughout North American history, disputes have been settled by each side hiring its own advocate to do battle in front of a neutral party. After the battle is concluded, the neutral then decides the outcome. At first glance, this seems to be a sensible process. Not only does each side receive a confidential evaluation of its position from a person it has hired as a loyal advisor, but that advisor also serves as a champion to assert the client's rights. After a process of information gathering and exchange, the neutral party (judge, jury or arbitration panel) makes a ruling based on what the advocates "introduce into evidence."

Very elegant and orderly—until you lift the lid and look inside.

Inside you will find black holes. The biggest of these is what attorneys call "discovery." This is the state-regulated procedure that occupies much of the time between the filing of a civil lawsuit and the trial. (Normally discovery is not optional—it is required.) On the surface, discovery is for the purpose of each side's learning about the case, but it goes much further than that. It is also an opportunity to identify the opponent's weaknesses through a variety of prescribed methods, among them document requests (obtaining copies of correspondence, notes, contracts and e-mails in the possession of the opponent or third parties), interrogatories and requests for admissions (asking written questions to the opponent which must be answered in writing), and depositions (interviewing the opponent's witnesses before a court reporter).

This can take heaps of time. In fact, a not-at-all-outrageous discovery timeline in an average case might look something like Figure 9.

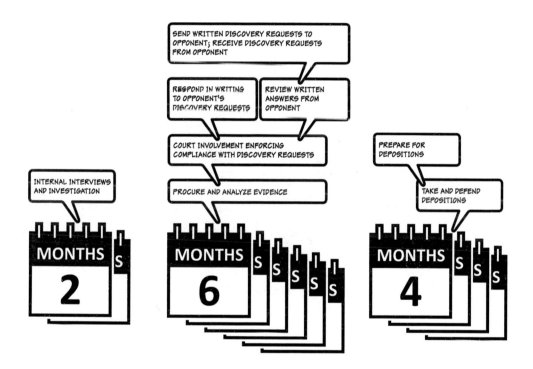

Figure 9—The American Civil Discovery Process.

This is just the gist of the discovery process—and modern technology has only made it even more complicated and costly. The seismic shift that has occurred over the past two decades as business and industry have wrenched themselves from the world of manila folders and file cabinets into the world of servers and gigabytes has been matched by one in the legal system. Not only are there brand new electronic discovery rules, there is a cadre of forensic computer analysts ready to seek out all of the deleted information that may be hidden in hard drives.

And discovery is just one of the big categories of ingredients that go into the litigation soup (Figure 10). There are at least two others of significance to technology companies: motions and experts.

Figure 10—The Key Ingredients in Litigation Soup.

Motions

Motions are what move the North American civil litigation process from beginning to resolution.

First, a very important digression: there is a common misconception among non-lawyers that judges and courts are continually "plugged in" to the lawsuits they oversee. To put it bluntly, they are not.

Client: We just filed a lawsuit in Judge Smith's court. This subject matter certainly should pique his interest.

Grizzled attorney: Actually, it won't.

Client: Why not?

Grizzled attorney: Well, because he won't actually read it.

*　　　　　　　　　　　* * **

Six months later ...

Client: The case sure has been dragging on. Judge Smith must be getting really frustrated with the lack of progress.

Grizzled attorney: Actually, he isn't.

Client: Why not?

Grizzled attorney: Because he never thinks about this case.

While it may seem strange to hear—and, admittedly, contains an ounce or two of exaggeration for emphasis—the above conversation more or less states the reality. Courts in general, and the civil servant judges that populate them in particular, sit atop a mound of filed paper. Nearly all of that paper describes disputes that are of little or no significance except to the people who are involved—and which only rarely, if ever, are reported in the media. Indeed, except for the persons on the business end of the process server or sheriff, the filing of a lawsuit is like the proverbial falling of a tree in a forest without a witness to hear it. Although the papers are numbered and pigeonholed, and filing fees are collected, the assigned judge neither sees the papers nor evaluates them, and blissfully goes forward uninterrupted as if the whole thing never happened. In the absence of an advocate paid by the person who filed the lawsuit to move the process forward to the next step, the case will slowly, quietly shrivel up.

That's where motions come in. Motions are the vehicle by which litigants force the system to pay attention to their demands—and (in principle, at least) move the lawsuit forward to a conclusion. They are also the vehicle by which opposing attorneys challenge claims, erect barriers, fight back and (all too frequently) waste everyone's time.

Whenever a motion is filed (in writing), the judge "wakes up" because the system is now asking her for a decision. This does not necessarily mean that she then takes the entire file home and reads it from beginning to end—with hundreds of other cases demanding attention, there may be insufficient time for that. Instead she relies on the attorneys to tell her, often for the first time, who the parties are, what the case is about, and the short-term "relief" that is being requested (e.g., "force the other side to give us its e-mails" or "set a date for the disclosure of expert reports"). After the motion is filed, the opposing party is given an opportunity to respond, again in writing, and then the judge will typically invite the attorneys to argue the matter in person before rendering a decision. Between the preparation and filing of a motion, extensions of time to respond, response, oral argument at hearing, drafting the decision and finally its issuance, a *single motion* may use up more than a hundred days—and more than a hundred hours of attorney time.

In the time between the filing of a lawsuit and the day, many months or even years later, when a jury or judge finally reaches a decision on the ultimate issues, there are numerous opportunities for motions and related requests for court action. Figure 11 depicts the most typical.

Suffice it to say that any one of these requests for court action has the potential for involving considerable strategy and gamesmanship, written argument, travel, legal research—and, not unsurprisingly, cost.

TYPE OF MOTION	TIMING	ASSERTION	RESULT IF SUCCESSFUL
Motion to dismiss; demurrer; preliminary objection; motion to strike; motion for judgment on the pleadings	Immediately after filing of lawsuit	The law says we win even if everything the other side says is true.	Case or part of it ends with a win or loss.
Motion to stay	Immediately after filing of lawsuit	The case should be put on hold or handled in a different way.	Case is put on hold or referred to a different type of proceeding.
Motion for a continuance; motion for an extension of time	At various times	The time deadlines should be extended or the proceedings should be postponed.	Delay or postponement.
Motion to consolidate	Immediately after filing of lawsuit	The case should be combined with another case.	Cases are combined with each other.
Motion to compel	During discovery	The other side needs to be forced to cooperate with discovery.	Cooperation is ordered.
Motion for a protective order	During discovery	The other side's discovery should not be allowed.	Discovery in question is prohibited.
Motion for summary judgment; motion for judgment as a matter of law	At various times, but often after discovery	Now that the undisputed facts are known, there is no need for a trial—we win.	Case or part of it ends with a win or loss.
Motion seeking mediation	At various times	The parties should settle their dispute out of court.	Case is referred to a third-party mediator, who will conduct non-binding settlement talks (case returns to court if unsuccessful).
Motion for trial; praecipe for trial	After discovery	The case is ready to go to trial.	Trial date(s) set.
Motion in limine	Immediately before trial	Certain evidence should not be heard at trial.	Evidence is prohibited from being introduced.
Motion for judgment notwithstanding the verdict; motion for a new trial; motion to correct error; motion to reconsider	Following a verdict or ruling	A mistake needs to be corrected or the proceeding was flawed.	The result is changed or a new trial is ordered.

Figure 11—Types of Motions in American Civil Litigation.

Experts

Everything I have been discussing so far could be applicable to any type of civil litigation of any complexity—not just disputes involving automation providers or their customers. The final big "bucket"—experts—is somewhat different. The whole idea behind a court designating someone as an expert to testify regarding a specialized topic is quite simple: it is that the judge, juries and arbitrators often need help in understanding complex or technical subject matter.

As any automation professional knows who has struggled to respond in common English to the question "So what is it you do?," there are few subjects in which the legal system needs more expert assistance than with the technically intricate and acronym-heavy world of DCS, VFD and HMI. Thus, whenever a potential litigant is calculating—hopefully on the front end—the costs and trouble it will take to litigate a dispute, attention to the "expert problem" must always be given.

There are two types of experts that are of particular relevance to automation-related disputes. (Note: in practice they are often combined into a single person.)

The first is the purely technical expert. This is the person who explains, for example (ideally in simple, easy-to-understand language) what a system is, how it is designed, how it is supposed to work, the integration of code with devices, the methods of commissioning and troubleshooting that are involved—and inevitably, exactly what happened in the particular project in question, what likely went wrong, how it could all be made right, and the associated costs of all the above.

The second is what I will refer to as the duty expert. This is an expert (every bit as technically capable as the first type) who takes it a step farther and crosses over into the territory of law and lawyering.

In case you did not know, it is not enough to go to court to allege that errors have been made resulting in damage—there must also be some legal wrong that has been committed, whether breach of contract or breach of professional duty. In essence there is a professional measuring stick against which the performance of the target must be judged—and the duty expert is there to assist the judge, arbitrator or jury in doing the measuring. (For more on this measuring, see Chapter 15 *Negligence*.)

Although either type of expert is frequently a third-party hired gun—a person who makes his or her living giving expert testimony—this is not a requirement, and it is just as common for an employee or principal of the company that is a party to the lawsuit to be given that designation. Of course, the most important thing for any expert is credibility and conventional wisdom holds that a third party or outside expert appears to the court or arbitration panel to be more impartial than someone who was involved first-hand in the dispute being litigated. Even putting aside the fact that this supposedly impartial person is being *paid* for that testimony (a fact that always is emphasized by the opposing lawyer in cross-examination) this probably is an accurate appraisal. The reason the professional testifiers are professionals is that they know the ropes and do a good job. At the same time, the tremendous cost savings from using the testimony of an internal expert (especially if that person is particularly impressive and knowledgeable) should not be discounted.

CHECKLIST: Four Types of Proceedings

✔ **Litigation.** A term used to refer to all methods of resolving disputes in general and the government-regulated civil system for resolving disputes in court in particular.

- **Jury Trial.** Submitting a dispute to six or more appointed laypersons, who, after listening to testimony and argument from all sides, resolve the dispute based on legal rules supplied by the judge.
- **Bench Trial.** Submitting a dispute to a judge for decision. Some types of disputes must be decided by judges instead of juries.
- **Appeal.** Taking a dispute to a higher court following the initial resolution by judge or jury. Appeals allege that the initial decision was in error and should be re-decided or changed.

✔ **Administrative Proceeding.** The process by which a government department or agency issues a ruling on a matter over which it has responsibility. A hearing officer or administrative law judge typically decides the issue after reviewing evidence from those who are affected. The decision can usually be appealed by filing a lawsuit in a regular court.

✔ **Arbitration.** A parallel, private system for resolving disputes. By law, persons or companies may agree by contract to give up their right to resolve a dispute in a government-established court and submit it to one or more private arbitrators (also called neutrals) for a binding decision. This alternative system is designed to be quicker and less expensive (although this is not uniformly the case)—and there is no appeal.

✔ **Mediation.** A non-binding process, sometimes required by courts or contract as a first step before litigation or arbitration, by which a neutral third party attempts to broker the settlement of a dispute—typically at a cloistered meeting called for that purpose.

CHECKLIST: Three Indispensable Inquiries to Make Before Litigating

✔ **What are the facts?** Think of a pair of columns. In the left column go the good facts: They delayed us. They dramatically changed the project parameters in midstream. They lost their original project manager. The general contractor did not know what it was doing. In the right hand column go the bad facts: Our code was flawed. We had trouble manning the job. Our supplier was delayed. In the middle column go the important facts that favor neither side. (Important note: Will your managerial view of the facts survive a detailed review of the project file and your employees' written communications?)

✔ **What is the law?** What do the applicable contracts say? What does the law say once the facts are reviewed against the contracts? Do we (or they) have lien or bond rights? Do we (or they) have a right to reimbursement of our attorneys' fees?

✔ **What are the economics?** How much is at stake? What are the odds of success? What are the hard expenses (lawyer fees and expert charges) and the soft expenses (disruption to our business) of going forward? Is litigation really worth it? (Important note: It may be the case that the economics do not even justify looking very deeply into the facts and the law—sufficient to know whether the claim should be pursued or not!)

CHECKLIST: Eight Rules for Resolving Contract Disputes

Sometimes the meaning of a contract is unclear. If so, here are some of the rules used by courts to ascertain the correct meaning:

✔ A contract will be interpreted according to its plain meaning. Only if the meaning of a provision is *not* plain and unambiguous do any of these rules come into play.

✔ If there is a possible interpretation under which two otherwise contradictory provisions can be made consistent with each other, that is the interpretation that should be adopted.

✔ Look to any "order of priority" provision in the contract for resolving contradictions between separate contract documents—for instance between drawings and specs.

✔ If there are two provisions on the same topic, and one is general and the other is more specific, the more specific provision will take precedence.

✔ Handwritten or marked-up additions to a contract carry greater weight than pre-printed provisions.

✔ If there are two equally plausible interpretations of a contract term, the one most favorable to the company that did NOT draft the contract should be adopted.

✔ In the absence of being able to apply any other rule, the interpretation that the parties themselves have given the contract (either in this project or in past projects) is given great weight.

✔ When there is no other means of resolving confusion, custom in the industry will be used to decide the issue.

TIP: Beware of Making False Claims

Can a contractor legitimately include some "padding" in a claim on a project for intended use as fodder in negotiations with the owner? Not if the owner is the federal government (and some states). The federal government, and some states, have shown an increasing willingness to utilize the False Claims Act against contractors. It is important to keep in mind that, under the FCA, a claim can take many forms and can potentially give rise to liability in ways one might not expect. A claim can include a request for payment, for example, a monthly pay application. If the application misstates a fact and a contractor knows it, this is a false claim. If, for example, the contract overstates the amount of work done, or requests payment for work that does not meet the contract requirements (e.g., the contractor uses only three screws to secure a stud instead of the four required by the specifications), this can be a false claim.

What if a contractor fails to pay a subcontractor, but certifies in its pay application that it has? If the contractor does not inform the government and return the money, and later submits an application in which it certifies that it has paid all sums received for a subcontractor, this and all future applications would be independent false claims.

Another dangerous area is the preparation and submission of claims for equitable adjustments, particularly claims in excess of $100,000, which must be certified. For instance, a contractor was found to have violated the FCA when it decided to set up a separate account for extra work and proceeded to direct charge all supervision and labor to the account, leaving no cost to be charged to the concurrent unchanged work. Even though the accounting practice reasonably supported the amount of money claimed for the extra work, the practice of charging all supervision and labor to the special account was found to be clearly designed to overstate the claim and therefore be false.

Chapter 15

Negligence

"Negligence" is a word that is thrown around whenever your company does not have a contract with the person or company that is asserting a claim against you.

When I use the word contract, I am using it in its broadest sense; that is, anything that has the flavor of an agreement—whether it is a formal written document with the word "contract" at the top and signatures at the bottom, a proposal followed by a purchase order, or a handshake deal. Whenever things break down and people are suing each other—and there is a contract—it is the terms and conditions of the contract itself that are under discussion. So you say you are suing my company because we allegedly did not fulfill our responsibilities? OK, then let's take a look at what those responsibilities actually are…

But with negligence we are talking about something else. There might be no terms. There might be no conditions. There might be absolutely nothing that could be construed as a contract. In fact, before the lawsuit your company might not have even known about the *existence* of the person or company suing you. Yet you may *still* be held liable in a court of law.

How does this liability-to-a-complete-stranger thing work?

Think of negligence as a deal you have made with society in general—a deal that covers not just those with whom you have made a contract, but potentially a great many others in the big, wide world. Those who are suing you are alleging that your company owed a *legal duty* to use *reasonable care* in carrying out your business activities. And they are alleging that they are exactly the type of person or company to whom that legal duty was owed.

How is a company to know whether a legal duty is owed or not? Unfortunately, there is not an easy answer to that question. That is because a legal duty may derive from a variety of different sources—sometimes limited only by the ingenuity of the lawyer who is doing the looking—such as:

- **Laws on the books.** A legislature or government agency may have passed a law or regulation that establishes that companies in your particular position or circumstance owe a duty to certain other persons in a related position or circumstance.

- **Common law.** Even if the legislature may not have given the matter specific attention, there may be such a duty that can be implied from years, decades—and even centuries—of written judicial decisions. This is where the resourcefulness of the lawyer is most in demand.

- **Contract**. Finally, the duty may be written into a contract. (Wait—I can hear you objecting: "I thought you said at the beginning of this chapter that negligence is the word that is thrown around whenever your company does *not* have a contract with the person or company that is asserting a claim against you." Yes, you have that right—I did say that. However, what I meant was that the lack of a contract can breed a claim of negligence because there may be no other claim. However, that does not mean (in many cases) that you cannot be liable under *both* a claim of contract breach and as a result of negligence. Think of the two concepts as subject areas with some degree of overlap, as depicted in Figure 12.)

Figure 12—Liability for Negligence and Breach of Contract Can Overlap.

For instance, a contract term may specify that the work of an automation provider shall meet the "highest applicable standards of the automation industry." In this instance, the provider finds itself straddling the worlds of contracts and negligence. (Put aside for a moment the sheer difficulty of determining exactly what those "highest standards" are. After thousands of dollars of lawyers and experts, the answer might be known. Far cheaper would have been to specify a particular standard, such as ISA-95.)

Perhaps a good way to understand the ways in which the concept of negligence menaces the work of the automation provider is to pause and reflect on one of the worst legal meltdowns affecting a control system that made it through the courts in 2007—a dozen-year affair that came to a grinding halt (in all ways that you can imagine that phrase) before the Mississippi Supreme Court.

Note to owners of control engineering companies: Any resemblance between these events and your worst nightmare is not coincidental. (Additional note to readers: Because I am interested in the educational—not the sensational—aspects of this, I am either not identifying or using pseudonyms for the companies or persons that were involved.)

Our story starts in 1995. A utility's power generating unit in Mississippi was being upgraded. The utility contracted with a plumber (no, I am not making that up) for the upgrade of its control system. The plumber subcontracted that work to a control system engineering firm, Best Integration Services (BIS).

The day came for site acceptance testing. Dave, the BIS control systems specialist, tried to start the turbine and bring it to its rated speed of 4,860 RPM, but the mechanical overspeed mechanism tripped and shut it down—even though the readout showed only 4,000 RPM and not the trip speed of 5,346 RPM. Dave then went to work trying to figure out what happened. Because he was by himself, he enlisted Bill, an employee of the utility, to test the speed of the turbine using a Strobotac. Bill measured the turbine speed with the Strobotac and called out the readings to Dave in the control room. The readings confirmed that the BIS control system was accurately measuring the speed of the turbine. So what did they do? You guessed it—they figured out the problem was the mechanical overspeed mechanism and adjusted it upward. When the turbine was started up again it ran, but vibrated excessively.

Two months later the real problem was discovered. Using a digital tachometer, it was determined that the turbine was actually operating at a speed of 6,560 RPM, although the BIS control system was only showing 4,860 RPM. Apparently BIS had mistakenly placed its sensors on an auxiliary turbine shaft that moved at a different speed than the main shaft—and when the manual Strobotac readings were taken, apparently the wrong shaft was again checked! The result of the adjustments made in reliance on these measurements was the destruction of the turbine rotor, followed by 12 years of litigation.

The final chapter of this story was written in 2007 with an opinion handed down by the Mississippi Supreme Court. As you might suspect, there was no happy ending for the control system engineering firm, BIS. BIS and the plumber were ordered to pay $2.5 million in damages, including repair costs, interest and attorneys' fees.

Each of the various defenses of the engineer was shot down. For instance (and I am paraphrasing):

> **Engineer**: It was the owner's employee who took the erroneous Strobotac readings, not us. So the owner should bear at least part of the blame.

> **Court**: The contracts allowed for the owner to assist the engineer, but that did not change the fact that all of the legal responsibility for the control system remained with the engineer.

> **Engineer**: There is no evidence that the owner's employee did not measure the speed of the main shaft.

> **Court**: Even without direct evidence, it is only logical to assume that the employee would have been instructed to measure the shaft with the sensors—in other ones, the wrong one. But so what? The engineer was still in charge and legally responsible.

Perhaps the only ruling to go in favor of the engineer was that the utility was not entitled to its consequential damages (in this case, the utility's costs of purchasing additional capacity to make up for the loss of the turbine). This was not because such damages were out of bounds; it was because the court decided that the utility had not done enough to prevent these losses from happening. Which means the $2.5 million could have been even worse.

Yes, this was a perfect storm of things going wrong for the control systems engineer. But could this happen to anyone? In my view, yes—which is my not-so-subtle message regarding all things legal in the automation field. Whenever you cause a significant loss to someone else, you can count on the victim making the argument that it was your duty to prevent that from happening—regardless of whether it really was.

CHECKLIST: Three Critical Ways to "Manage" Negligence

- ✔ **Process.** "Process" means nothing more than giving careful attention to your company's ways of doing things. It is the way employees are trained. It is how often they are re-trained. It is the way they are managed. It describes the tests, checks and balances that are institutionalized. It is the safety culture. Promoting a better process may not prevent negligence in all circumstances, but it is highly likely to minimize the impact.

- ✔ **Contract.** When there is no contract, the negligence virus is unrestrained. The alleged "duty" of the negligent person will be entirely decided by outsiders—perhaps by standards and common law rules no one realized were applicable. When there is a contract, this situation is improved (although not entirely eliminated). A contract can define ahead of time the rules of negligence between the parties that sign it and can even divvy up responsibility for claims of negligence by third parties.

- ✔ **Insurance.** One truly cannot manage or mitigate liability for negligence without a thoughtful insurance strategy. Insurance is not just helpful in this arena; negligence is the reason somebody invented insurance in the first place. For the different types of insurance relevant to automation, see the discussion in the next chapter.

TIP: Who Is Responsible for Functional Safety?

The word "safety" means taking steps to prevent personal injury. It should be no surprise that safety concerns both negligence (this chapter) and contract (earlier chapters). Negligence is implicated because it usually is someone's duty to prevent injury—thus, whenever someone is injured the victim predictably claims that a duty was breached. There are at least three different, but related, types of common law safety duties:

- The duty to eliminate the safety hazard altogether.
- The duty to protect against that safety hazard (if it cannot be eliminated).
- The duty to warn about that safety hazard (if it cannot be protected against).

In addition to these common law duties (which can be unrelated to any contract), most automation contracts require at least some attention to safety—not to mention indemnification against third-party negligence claims. Thus, whenever a person is injured, a breach of contract or indemnity claim also often results.

In the industrial automation realm, safety receives particular attention because of the obvious capability of some types of moving machinery to injure humans. So, when an automation-related injury occurs, companies at each stage of the process frequently are targeted because each arguably had the opportunity to prevent the injury from happening.

An orderly way of analyzing the responsibility for machine and process control safety is to ask a series of questions:

1. What safety obligations, if any, are imposed by law? (This includes OSHA regulations, both the federal ones and the state equivalents).

2. What safety obligations, if any, are imposed by contract? (Paradoxically, the more a contract lays out in detail the duties to prevent injury, the more easily an injured person can argue that those obligations were not observed.)

3. What safety obligations, if any, are imposed by custom, best practice or standard—without regard for any contract? The existence of such obligations, even if never mentioned or observed in connection with the contract or project in question, can be evidence that "reasonable care" was not followed, prompting a claim of negligence. Among the standards, terms and practices potentially applicable to automation projects are (a) IEC 61508 and 61511, each standards relating to the safety life cycle of industrial systems, with 61511 zeroing in on a type of instrumented safety system used within the process industry called the Safety Instrumented System (SIS) and (b) the Process Hazard Analysis (PHA), which is a methodical assessment of the hazards of an industrial process using a variety of proven techniques.

Chapter 16

Insurance

Among the very first inquiries that should be made whenever a "business mess" materializes is whether the problem is insured. (Of course it goes without saying that insurance coverage should also be addressed before ever going into business—let alone starting on a project—but let's put that issue aside for the moment.)

For what it's worth, my own threshold for what constitutes a "business mess" is fairly simple: Is this problem either very expensive or potentially very expensive? In other words, I am not talking about those issues that are frustrating or disheartening (or even somewhat expensive) but at the end of the day *manageable without outside help*. (Question: Are you really going to notify your insurance company when a control panel is dented as a result of dropping a tool? If you are sensible, you would not. So we are talking about the big stuff.)

In response to the question "Is this business mess insured?" there are two, somewhat (but not entirely) facetious answers:

The first answer is that **no business mess is insured**. This, with an ounce of exaggeration, is the position of the insurance industry. The insurance companies will tell you, "We insure accidents, not business risks. An unfavorable business result, even one that rises to the level of a 'mess,' is not an accident." Thus, if your company was depending on a particular profit margin on a project and ended up losing money, you would not expect to make a claim for the loss from your insurer. No insurance policy covers business risk, right?

But the second answer—you guessed it—is that **all business messes are insured**. This is the position of, well, any policyholder who is motivated enough to take the time to parse the applicable policy. Reaching this place starts with the realization that an insurance policy is never really negotiated; it is a "take it or leave it" proposition. This, as it turns out, is actually an advantage for those who challenge an insurance company's coverage position because, as far as the courts who interpret these policies are concerned, any ambiguity in meaning is resolved in favor of the policyholder.

Those are the (admittedly exaggerated) extremes that define the territory. Now for the practical reality:

Any analysis of whether a loss is covered under an insurance policy consists of giving attention to three distinct parts of the document (including endorsements): (a) the coverage grant (sometimes called the "insuring agreement"), (b) any exclusions to the coverage grant and (c) any exceptions to the exclusions.

The first step is to examine the "coverage grant." The coverage grant is the set of "big picture" sentences (often at the beginning, but not always) that lay out the overall intent of the insurance policy, typically in broad terms. It *gives* coverage to the policyholder.

The second step in this analysis is to examine what is typically the largest part of most insurance policies: a long list of exclusions. This is a list of those matters that, although ostensibly within the coverage grant, are not covered—these provisions *take away* coverage.

Third, examine the exceptions. Typically embedded within each exclusion is one or more exceptions (think of these as sentences using the word *unless*). They *add back* coverage. Whatever remains is the covered loss. Think of the basic scheme as an umbrella with holes (and patches in some of the holes), as pictured in Figure 13.

So, putting all of this together, a fictional policy to cover injuries caused by dog bites might read as follows: "This policy covers injuries you received from dog bites (the coverage grant). It does not cover being bitten by dogs owned by you (exclusion) unless your dog bit you as a result of being excited by a dog not owned by you (exception to exclusion)."

With the above (overly simplified) description in mind, let's now take a look at the most common types of insurance coverage purchased by companies doing business in the automation and process industry field:

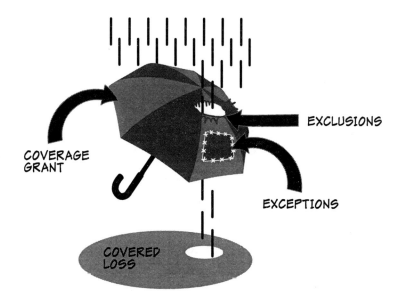

Figure 13—The "Umbrella" of Insurance Coverage.

Commercial General Liability Insurance

Commercial general liability coverage (often abbreviated as GL or CGL coverage) is the "big dog" of the realm—the type of insurance that most automation-related companies correctly decide they cannot be without. These policies insure against liability for bodily injury (or death) or property damage caused to others. An example of a loss resulting in property damage in the industrial setting might be the following:

- An error in programming results in an unpredicted movement by a robot that destroys expensive peripheral equipment.

- The owner of the facility makes a claim against the company that programmed the robot.

- The company that programmed the robot makes a claim on its commercial general liability policy.

Based on the wording of standard policies, there is little doubt that in this situation the CGL insurance policy would respond to the claim. This means that if a lawsuit were to be filed, the insurer would hire lawyers to defend against the lawsuit and that, to the extent the lawsuit is settled or lost, would pay damages.

(Even if a lawsuit is not filed, the insurance company may still write a check to the extent the party suffering a loss makes a demand for reimbursement.)

What seems fairly straightforward becomes thornier, however, as additional complexities are introduced. Let's consider the same basic example, with a few changes:

- An error in programming results in an unpredicted movement by a robot that destroys expensive peripheral equipment.

- The owner of the facility files suit against the company that installed the robot. The owner of the facility requires the company to replace the robot (and programming) at its own cost.

- Now, let's assume the installation company had subcontracted the robotic programming to another company, but the company that installed the robot is unable to make a claim against the subcontractor because the subcontractor is out of business.

- The company that installed the robot makes a claim on its commercial general liability policy for all of its out of pocket costs, including the property damage loss and the cost of replacing the robot.

With these additional complexities the fine print of the insurance policy springs to life. While there certainly is "property damage" that ostensibly brings the loss within the sphere of CGL coverage, part of the alleged loss is the cost of replacing the robotic system itself. Is this loss covered by the policy? CGL carriers would typically say it is not, pointing to a standard exclusion that prohibits insurance proceeds from paying for replacement of the insured's own work (lawyers and insurance professionals call this the "your work" exclusion). The justification for this exclusion is that the cost of pulling out and replacing the insured's own work is not really the sort of unexpected accident for which insurance is purchased, but instead something more properly addressed through the applicable contract between the parties—for example, the warranty provisions.

But remember what I said about exclusions taking away coverage, while there may be exceptions to the exclusions that give it right back? The hypothetical described above depicts not just an exclusion, but a possible exception. Because it was not the insured company itself that caused the damage (instead, it was a programming error by a subcontractor), a typical exception might be triggered by this occurrence. In standard CGL policies, the "your work" exclusion does not apply if the property damage was caused by a subcontractor.

Summing up, the logic of this grant/exclusion/exception formula might be described as depicted in Figure 14.

This policy covers claims alleging that you caused property damage **−** But not damage to your own work **+** Unless your work was damaged as a result of something your subcontractor did

Figure 14—The Logic of the Subcontractor Exception.

The reason I suggest the subcontractor exception *might* be implicated"—as opposed to *would* be—is because every U.S. state has its own "spin" regarding the purposes of such insurance and how the language of these provisions is most accurately interpreted. Thus, whether you have coverage for a particular business mess likely depends as much on the state in which the project was located as the wording of the insurance policy in question.

Professional Liability Insurance

Another important type of insurance policy in the automation realm is one that provides professional liability coverage (PL, otherwise known as errors and omissions, or E&O coverage). Unlike the type of protection provided by CGL policies, this coverage is not focused on a type of result (property damage or personal injury) but on a cause: negligence. For PL coverage to kick in there must be a breach of a professional duty resulting in a loss to others—which means, of course, that there must be such a duty owed in the first place (for more on how such a duty arises, see Chapter 15).

As with CGL policies, there is a formidable gauntlet of exclusions and exceptions. One of the most dissected PL exclusions is the one that rules out coverage in the event of so-called "contractual liability"—liability that is assumed by the insured under a contract. However, this is a much more complicated topic than one might immediately think; for instance, despite what I just said this exclusion does *not* mean that PL coverage is not provided whenever an automation provider is sued by a company with which it has contracted. What it *does* mean is that if there would be no liability on the part of the automation firm *in the absence of a contract,* the PL coverage is not triggered. To illustrate (using a version of the previous example), consider the following:

- A robot programmer has a contract with a facility owner.

- An error in programming results in an unpredicted movement by a robot that damages expensive peripheral equipment *owned by a third party.* The facility owner suffers no loss.

- The owner of the damaged peripheral equipment files suit against the facility owner.

- The contract between the facility owner and the robot programmer requires the programmer to indemnify the owner against claims by third parties arising from errors by the programmer.

- The facility owner asserts a claim for indemnification against the robot programmer.

- The robot programmer makes a claim against its professional liability policy.

Here, the so-called "contractual liability exclusion" might preclude the PL policy from responding. This is because the owner, *which did not suffer a loss directly,* would not have a direct claim against the programmer for negligence in the absence of the indemnification provisions of its contract with the programmer.

If you are thinking that there is overlap between the types of claims covered by CGL policies and those covered by PL policies, you are not missing a beat. Often there is—leading to two important lessons.

One lesson is that following a loss or claim, it may make sense to give notice to all potentially responsible insurance carriers, even if, based on the facts known at the time, one type of coverage and not another appears responsive. A second lesson, which you will hear from brokers, is that there can be an advantage to placing more than one type of coverage with a single carrier to avoid the finger-pointing between policies that sometimes results.

CHECKLIST: Nine Additional Insurance Coverages and Concepts Worth Knowing

While a comprehensive analysis of insurance policies is beyond the scope of this handbook, even a passing treatment of automation risk would not be complete without at least a basic discussion of these additional coverages and concepts.

- ✔ **Products and completed operations coverage.** It is one thing to have insurance coverage for a loss that occurs in connection with an ongoing project—it is another for the coverage to be there years after the project is completed. That is the function of what is called "products and completed operations" coverage. Although this coverage is normally included in standard CGL policies, there are plenty of catches. One is that for this coverage to be valid the policy must be kept in effect (i.e., premiums must be paid) for the entire extended period. Second, this component of CGL coverage does not necessarily transfer to a third party "additional insured" when purchasing the "additional insured" endorsements that are required on many projects.

- ✔ **Products coverage.** Many automation projects arguably include delivery of a "product." While for most of those projects the "products and completed operations coverage" often included in a CGL policy will be adequate, at the point when the deliverable becomes more "product" than "project," it may be wise to procure a separate product-focused policy. Such a policy will address some risks not covered by the general liability protection—for instance, the devastating economic consequences of a product recall.

- ✔ **Pollution coverage.** Because the release of dangerous chemicals within a processing facility can be a significant risk to automation companies that work in such areas, it is important to understand what types of pollution events are and are not covered by most standard CGL policies. The following types of risks are among those that usually fall *within* the scope of CGL coverage (subject to standard warnings regarding interpretive "gray areas" and the need for consultation with an insurance professional):

    - Liability for causing the release of pollutants that the insured company *not* responsible for bringing to an outside work site.

    - Liability for gases, fumes or vapors from pollutants that the insured company *was* responsible for bringing to an outside work site—to the extent that those fumes arise from the type of work that the insured company performs.

✔ **Builder's risk coverage.** Builder's risk is another must-have type of coverage on any project. It is, essentially, a specialized type of property insurance designed to protect structures or production areas while they are being constructed or renovated. Typically, but not always, it is the owner of the project that carries this insurance—but those working on the project should be equally motivated to make sure that *someone* is carrying this coverage to avoid an uninsured catastrophe. The risk of fire is first on the list of perils that are covered. Unlike CGL coverage, which insures against third-party claims, a builder's risk policy—like all property coverage—is designed to cover "first party" losses (which means that the claim is made directly by the company that bought the insurance). Builder's risk coverage gives way to "normal" property coverage following project completion. Because there can be multiple causes and consequences associated with any insured loss (defective work, negligence, third-party injury, etc.), the types of claims triggering a builder's risk claim can also trigger CGL or PL claims.

✔ **Workers' compensation coverage.** Workers' compensation insurance is another necessary type of coverage—not just because it is advisable, but because it is compulsory under the laws of all U.S. states. This insurance covers the medical expenses and lost wages of employees who may be injured during a project. The laws compelling this insurance put in place a sort of trade-off. The companies that are required to purchase it are (in return) protected against claims for negligence by their own employees. In the automation arena, where it is common practice to "borrow" or bring in temporary independent contractors to work on a system (often as a result of a demand by the project owner), this can produce complexities, with the result depending on the definition of "employee"—and whether the parties want to embrace or avoid the workers' compensation "shield." Because every U.S. state treats these issues somewhat differently, careful attention to the laws of each jurisdiction is important.

- ✔ **Wrap-ups, OCIPs and CCIPs.** Employing the principle of "economies of scale" owners or general contractors sometimes require key participants in a project to be covered under a common insurance program. Known generically as a "wrap-up," such a program is called an OCIP (owner-controlled insurance program) when the owner purchases the insurance and a CCIP (contractor-controlled insurance program) when the general contractor or construction manager makes the purchase. A few other things to know:
    - Typical coverages within a wrap-up include CGL, workers' compensation, builder's risk and pollution coverages. Professional liability coverage is typically not part of the program.
    - To capture the cost savings from avoiding duplicative insurance policies, participants in a wrap-up are sometimes required to submit two alternative bids, one that includes the cost of premiums for insurance outside a wrap-up if the wrap-up is not used and one without those premiums in the event it is used.
    - Wrap-ups are seen as only making sense on very large projects. Consideration of the advantages, disadvantages and numerous complexities argue for the early involvement of both legal counsel and skilled insurance professionals.

- ✔ **Additional insured endorsements.** A common provision in automation contracts requires the automation provider to make another party (sometimes more than one) an "additional insured." This means that the provider is being required to purchase insurance for the other party. The way this is typically done is via endorsement (an amendment to the original insurance policy) in which the carrier agrees, for the price of an additional premium, that the same coverage being given to the automation provider will also be given to the "additional insured." This practice leads to two common problems:
    - First, a provision in a contract or purchase order *requiring* such additional insured coverage is not the same thing as its procurement. All too often those who agree to provide such coverage do nothing to actually purchase it and many project owners are not diligent in following up to see if an additional insured endorsement was actually obtained. This failure can result in both an underinsured project and liability for breach of contract.
    - Second, even if it is successfully procured, the coverage provided through an additional insured endorsement is sometimes not at all identical to that provided by the underlying policy. Significant gaps may exist, again leading to big problems.

- ✔ **Waiver of subrogation.** One of the conditions of most insurance policies is the right of the carrier to step into the shoes of the policyholder—to be *subrogated* to the policyholder's rights—to the extent that it pays a claim. This means that if there is someone other than the policyholder who is liable for the loss, the carrier can pursue that someone in the place of the policyholder in an effort to be reimbursed for the payment (insurance companies call this "recoupment"). The inclusion of a "waiver of subrogation" provision in a contract shuts down that typical progression. In short, the policyholders are giving up the right of their respective carriers to seek such recoupment. The idea behind such pre-claim waivers (permitted by most insurance policies) is that they tend to reduce litigation. Waiver of subrogation clauses in the automation context usually apply only to losses covered by property insurance (including builder's risk) policies, not other types.

- ✔ **Defense and indemnity.** Whether or not the subject matter of a lawsuit is covered by insurance often is a matter of dispute. When that happens, the insurance company may notify the policyholder that while it is hiring lawyers to defend the claim, it is "reserving its rights" to deny the claim as being outside coverage. For this reason, it is often said that an insurance company's "duty to defend" (hire lawyers to protect its policyholder from claims) is greater than its "duty to indemnify" (pay claims).

Chapter 17

Liens, Bonds and Other Remedies

Whenever a payment dispute arises between an automation provider and a customer, at least two parties and one pot of money are involved. To illustrate the typical problem, let's give labels to these adversaries. Let's call the end user (or owner) the "have" and the automation provider the "have not"—meaning that the end user is holding money that the automation provider is claiming, as depicted in Figure 15.

Figure 15—A Two-Company Payment Dispute.

(Of course, I realize that many disputes are more complicated than that. For instance, the end user may also claim that it is owed money by the automation provider, or there may be two or three other companies that are involved in the claim. But let's keep it simple for the purposes of illustration.)

Here is the question: what if the end user (the "have") does not really have it? In other words, what if the pot of money (or moneybag or coffer or whatever image you prefer) is insufficient to cover the claim? That, boiled down to its basics, is where liens and bonds come in. Liens and bonds can be thought of as additional pots of money backing up the one held by the primary source (the target) that may be available to satisfy a claim.

In the case of liens—typically called mechanic's liens—that additional pot of money is the real estate (including building and improvements) where the work was done.

In the case of bonds—typically called surety bonds or payment bonds—that additional pot of money is a promise by a third party (called a surety) that if the primary source does not pay the claim, it will.

Liens

Liens are the creation of state legislatures in laws that date back to the 1800s. The age of these laws is the reason for their use of antiquated terms like mechanic, laborer or materialman in the place of the modern-day equivalents (contractor, subcontractor, supplier).

Mechanic's liens, because they are a remedy against real estate (and the buildings and other improvements upon real estate), are typically filed or recorded in the land records maintained by local governments—resulting in a claim against the particular tract or parcel where improvements have been made. While the requirements differ significantly between states, one outcome is a constant: the holder of the lien has the right to sell the real estate to pay the debt owed.

Needless to say, recording a lien can be a powerful inducement to payment.

Because liens are so powerful, the law has imposed a formidable collection of small print governing their use. This small print can quickly complicate (if not completely wipe out) the power of a lien. Here is a partial list of such complications:

Is the work really an improvement to the land? This is a problem specific to automation providers, which typically provide systems comprising not only equipment, but also software and the considerable know-how required to make it all work together. Some states limit the use of mechanic's liens to proj-

Figure 16—A Two-Company Payment Dispute with a Lien Claim Added.

ects resulting in a fixed improvement (i.e., a fixture) that is physically attached to the land. Providers of automation systems can be vulnerable to the allegation that their work does not result in such a fixed improvement, but is instead moveable or easily removed. Think about the difference between a slab of concrete and a cabinet of PLCs and you can see the problem.

Who owns the land? It may be that the owner of the real estate is not the same individual or company that signed the contract with the provider of the improvement. If that happens, some states require the owner to have either actively or indirectly consented to the improvement being made. This consent sometimes can be difficult to demonstrate. Related to this, it may be the case that the signer of the contract is really just the tenant under a lease—not the owner of the land. If so, it may be the case that only the lease interest—as opposed to the land ownership—can be captured with the lien (not an impossible task, but with unique complexities).

Public land. Liens are typically not permitted to be recorded against publicly-owned land, including water or sewer districts, airports, schools and public highways. To provide a substitute remedy, bonds may be required by state or federal law (see below for more regarding this remedy).

Are there other liens? The situation where a lien is most needed is the one in which the party owing the money does not have the ability to pay. But this often means that there can be a great many other creditors out there awaiting payment from the nonpayer—each of which may have its own ability to record a lien. Add to that one or more lenders with mortgages and there is now a crowd competing for whatever value that the land and improvements may have. Here is the important question: which among these has priority? In some states, holders of mechanic's liens are first in line for the money; in others, liens are equal to or lesser in priority to mortgages held by banks while in still others, it matters which claimant got to the land office first.

Have all of the Ts been crossed? There is also a gauntlet of purely technical requirements that must be satisfied to create liens. There are both notice requirements and hard deadlines. A typical notice requirement is that a subcontractor must provide notice to all potentially-affected parties within so many days of its *first* day of work (think of a letter that says, more or less: "Here I am. I am not recording a lien against your real estate, but I just might in the future. So make sure I am paid." A common (and usually inflexible) deadline is that the lien itself must be recorded or filed within so many days, weeks or months of the *last* day of work. Those are just the start. The description of the real estate must be accurate, the owners and other parties must be listed accurately and the amount claimed may not be overstated. In many states, all it takes is one strike and you are out, your claim null and void.

Is there a foreclosure? In many cases a lien by itself accomplishes nothing. Although it certainly cannot be ignored on the land records (shouting out for all to hear that Company X has a claim against Company Y's real estate) nothing really happens until a further step is taken: namely, asking a court to order the sale of the land in question. This is commonly called a foreclosure sale and it's more or less the same thing as the mortgage foreclosures you read about. To make this happen, the lien holder needs to actually file a lawsuit against the company that owes the money. This obviously costs money (which the lien holder might not have because it has not been paid). To get around this problem, many states force the target of the lien to pay the legal costs of the lien holder—assuming, of course, that the lien and underlying claim are ultimately ruled to be valid.

Dangers. It should not go unmentioned that because they are so potent, liens, like lawsuits, can also backfire, and end up opening up a Pandora's box of troubles. One danger is the possibility of "slandering title." This can occur if an invalid lien is maliciously or, in some cases, recklessly recorded against real estate, resulting in damage to the owner. For instance, when the owner is unable to obtain refinancing because of the presence of the lien, the lienholder may become liable for any loss that results.

Bonds

Bonds are a remedy in which a surety guarantees the **payment** (payment bond) or the **performance** (performance bond) of a contractor. Unlike liens, bonds do not provide a second pot of money—bonds instead represent a *substitute* pot of money that is only available if the prime source is not there. In fact, it would not be an exaggeration to observe that bonds are typically not needed unless a project truly has "gone south" and the party that promised payment is either out of business or is on its way to insolvency.

Payment bonds, which deal exclusively with the payment obligations of the "have," do not appear by magic. They are purchased on the open market because either a contract or a law requires it. It is a purchase with consequences. Not only must the "have" pay a premium for the bond, it must also make a promise. The promise by the "have" is essentially that the bond will never be called upon—and that if by some weird chance it is, the "have" will reimburse the surety.

That's why bonds are nothing like insurance. To quickly see the distinction, imagine an insurance policy where the insurance company pays for rebuilding your home after a fire, then has the ability to file a lawsuit against you to recover the money that was paid. Sureties are not the buddies of those who purchase the bonds—they are an unforgiving uncle of last resort.

In automation projects, payment bonds are typically purchased by the larger contractors who hold the contract directly with the owner, but the protection is really for others. One protected party, in a sense, is the owner, which naturally has an interest in making sure all of the downstream subcontractors and suppliers are paid. But the owner cannot make a claim on a payment bond. That is for those downstream subcontractors and suppliers themselves. If the "have" (the general contractor) does not pay, subcontractors and suppliers can seek to be paid by the bond company.

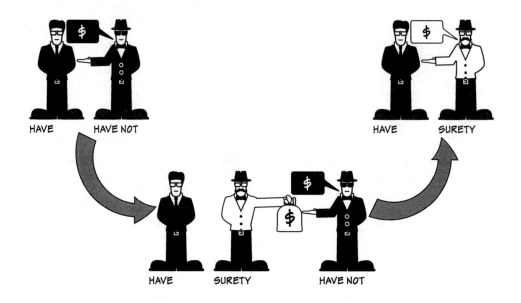

Figure 17—A Two-Company Payment Dispute with a Payment Bond Claim Added.

An important note: payment bonds are much more common on government projects than on private projects for a simple reason. It's because mechanic's liens—which you may recall give contractors the right to sell the land and improvements to pay what is owed—are not typically allowed on government land (imagine a subcontractor ending up owning a highway or a public building and you will see the problem). As a substitute for that important remedy, there are both federal and state laws that *require* general contractors to purchase payment bonds as a condition to doing work on a government project. The downside: there are strict deadlines for asserting bond claims on public projects that are similar to those found for mechanic's liens.

Performance bonds are similar, but also a bit more complicated. Here, the guarantee being purchased is the ability of an automation provider to follow through on what it committed to do on a project. If the principal (the company that purchased the bond) fails to live up to what it promised, the obligee (the company, person or agency to which the guarantee was given) can, in certain circumstances, call upon the surety to take its place.

This creates complications, for all of the reasons that you might guess. For one thing, whether the principal did or did not live up to its important performance obligations (with failure justifying making a claim against the bond) may be a matter of dispute. Related to this, the surety has various options for dealing with such a claim. It can deny the claim altogether—usually leading to lawsuits. It can agree to step into the shoes of the failed contractor and complete the project, often by hiring a replacement. Or it can simply pay the face amount of the bond to the person, company or agency making the claim.

Performance bonds serve another function that has nothing to do with their guarantee role. They also reflect a sort of corroboration of a company's balance sheet. Sureties—it should come as no surprise—are in the business of making money, and paying claim after claim is not the best way to stay in business. For that reason, the most reputable bonding companies (most of them owned by insurance companies) are only willing to sell their bonds to companies that are unlikely to get into trouble—namely, the best, most substantial enterprises out there. Owners therefore use this threshold ("Do you have the ability to get a bond?") as a test for determining whether a new contracting partner can be relied upon.

CHECKLIST: Six Remedies Other than Liens or Bonds

✔ **Impounding amounts due to upper tier contractors.** Many U.S. states permit a subcontractor or supplier to transmit a notice to the owner that it has not been paid by an upper tier contractor. To the extent that there is still money owed by the owner, the law then requires the owner to withhold payment from the upper tier contractor in the amount of the subcontractor's claim. Depending on the state, this right can exist on either private or public projects, or both.

✔ **Prompt payment laws.** Some U.S. states have laws on the books that require payment on certain types of projects within a set amount of time. When late payment occurs in these states, the matter is no longer a private issue but one in which the government has an interest.

✔ **Non-surety guarantees.** In a situation where a contracting partner seems sketchy, it can be a good practice to require a related party—say an affiliated company or an individual—to guarantee that partner's payment or performance in a separate written agreement.

✔ **Joint check agreements.** Similarly, if there seems to be a likelihood that your payment to a subcontracting partner is not making it through to a lower-tier subcontractor or supplier, it can be a good practice to have the right to make payments in the form of a check made out *jointly* to the contracting partner *and* the potential downstream claimant.

✔ **Security agreements.** When supplying expensive equipment to a questionable buyer, it is possible to guarantee payment by making the equipment security for the debt. When payment is not made, the equipment itself can be subject to reclamation or sale. The legal requirements for establishing such a secured relationship are not insignificant, but the protection is strong.

✔ **Demand for adequate assurance.** Under laws existing in nearly every state, parties to certain types of commercial supply contracts have the right to demand "adequate assurance" from a contracting partner if that party has "reasonable grounds for insecurity" that the contracting partner can pay or perform. If the assurance is not given, the company making the demand can be relieved from further contractual obligations.

Chapter 18

Maintenance and Service Agreements

The system has been commissioned. Demobilization is underway. And yet the end user and automation provider are not quite ready to part ways. How is the maintenance and service relationship best "papered" (made formal)? Is it ideally left to a handshake, or is it better to establish a long-term contract?

Let's approach the problem by defining what is meant by "maintenance and service agreement." (In doing so, let's assume first of all that a design and/or installation relationship between the end user and integrator preceded the service work—as is often the case.) An acceptable definition might be the following:

> *An agreement to do **additional work** to **maintain** or provide **support** to a **system** for **additional pay** above and beyond the original design / installation work.*

The key terms are highlighted. First, we are talking about work that truly is additional. It is not work that would have been done anyway in connection with the original installation. Second, the work must maintain or support a particular system (or set of systems). This means that work to retrofit, upgrade, augment or revamp the existing functionality is outside the scope. Third, there must be additional pay connected to this work. The best way of knowing that something stands on its own as an obligation (and as independent scope) is the willingness of someone else to pay for it.

Now let's engage in a bit of classification. As I see it, there are three types of maintenance and service agreements. The first is an alternate within the original design and installation agreement. Check the box, pay this amount and the "maintenance and service" scope is added. The second is via an independent written contract (the best choice, if you were polling attorneys). The third option—and far and away the most common—simply consists of the provider's picking up the phone, listening to the immediate problem and taking action whenever appropriate (we'll call this a "callback" relationship). The agreement? There is not much of one—"You are the customer, your satisfaction is important to us and we will try to get the thing to do what you want it to do." Never mind that there may be some muddiness as to whether this is original scope, a warranty call, a retrofit, an upgrade, or maintenance for a fee.

Each of these arrangements has its strengths and weaknesses from the perspective of the automation provider, as reflected in Figure 18. The "alternate to the original contract" approach is clearly the one most easily sold to the customer. The provider is already there in the building, so to speak, and it certainly makes the most sense to purchase maintenance and service protection when the project is in its infancy, before the matter can be mixed up with the inevitable completion problems. On the other hand, this is also the moment when the economic realities of the system are perhaps least understood, and pricing the services so that they are profitable (from the provider's vantage point at least) can be challenging.

TYPE	SALABILITY	LIABILITY	PROFITABILITY
Alternate in original contract	Strongest	Moderate	Weakest
Stand-alone contract	Moderate	Strongest	Moderate
Callback	Weakest	Weakest	Strongest

Figure 18—Strengths and Weaknesses of Maintenance Agreement Types from Automation Provider Perspective.

At the other end of the spectrum, the callback approach is naturally the most profitable because it is, by its nature, time-and-material based (we do the work and you pay whatever it costs), but there are some negative liability consequences. Among other things, the arrangement depends on unwritten (and therefore uncertain) agreements, there is no limitation of liability, and the dividing line between original scope, warranty, retrofit, upgrade and maintenance is obscure at best—which increases the likelihood of disputes.

That is why in many circumstances, the best arrangement for both user and provider of automation services is a stand-alone contract entered into either somewhere between the midpoint and substantial completion of a project or independent of a project altogether. From the provider's perspective, it is moderately saleable and profitable (most importantly, the provider is in a position to assess the system in question) and it is grounded in a detailed written agreement that specifically addresses the post-completion environment.

First-rate maintenance and service agreements are not off-the-shelf, but are tailored to the application in question—rarely will a generic service agreement do the job well. The must-have legal clauses fall into two categories: those that are specific to the maintenance and service relationship and those that are not.

The sections specific to maintenance and service contracts deal with payment, definitions, warranties, exclusions and what I call response boundaries. Those sections of a general character (call these "the usual suspects") address limitations on liability, termination, confidentiality, force majeure, disputes and completeness. The former set will be discussed here. The latter are discussed in Chapters 3 through 6.

Payment

The payment terms available for maintenance and service agreements are superficially the same as those for automation contracts in general—lump sum, guaranteed maximum and cost plus—with two important forewarnings. The first is that there is inherently an open-ended quality to the scope of these agreements that argues for time and material billing. The second is a competing dynamic—the fact that end users (like car owners) view the maintenance and service relationship as a sort of protective shield, arguing for closure (pay this fee and you do not need to worry any more). Given these opposing forces, the guaranteed maximum (or some variation on that theme) or lump sum with an hours threshold and fixed rates are acceptable compromises.

Definitions

Certain contractual terms cannot be left to themselves in a successful maintenance and service relationship, among them the definition of "system," the precise extent of the "covered system" in question, the boundaries between "customized" and "third-party" software, and the scope and nature of the services themselves.

Warranties

Of all the lines between original scope (design/installation) and maintenance/service, none is more fraught with difficulty than the line involving warranties. No end user wants to be invoiced for that which would be free in the absence of an extended service agreement; no integrator wants to be denied payment for additional work because the end user is unfairly raising the warranty flag.

Exclusions

Related to this is the need to carve out what is *not* covered by the maintenance and service agreement. Services provided to someone other than the end user certainly are not. So is migration from one platform to another. Upgrades are either covered or not covered depending on the system and the type of upgrade in question. It also may be wise to exclude the following:

- Services arising from an end user's failure to maintain the installation and operating environment in accordance with system documentation.

- Services arising from damage to the system or any of its parts beyond ordinary wear and tear, including accidents; unusual physical, electrical or electromagnetic stress; neglect; misuse; failure or fluctuation of electric power, air conditioning or humidity control; excessive heating; natural disasters such as fire, flood, wind, earthquake or lightning accident; or fire and smoke damage.

- Services that involve deviating from the best practices of the control system industry.

Response Boundaries

Another type of boundary is the one involving responsiveness to end user needs. The maintenance relationship inherently has a response time component that is minimized if the integrator has a facility presence but is maximized if the integrator has none. How quickly is a response required? *It depends* is the answer. One way of structuring this facet of the relationship is to define degrees of magnitude right there in the agreement (defining what is meant by an emergency and a wish list item and calibrating the divide between them). Thus, on one end of the spectrum, an emergency may require a substantive response within, say, two hours, while the least pressing matters may have to be addressed within two days. (See Figure 19 for an example of one way of structuring such an arrangement using defined severity levels.) To forestall abuse by either side, incentives and penalties for mischaracterizing problems can be built into the arrangement.

SEVERITY	DEFINITION	RESPONSE TIME
Severity 1	Business stopping outage; total enterprise-wide or mission-critical application failure.	2 hrs
Severity 2	Business function is operable but substantially degraded; affected area(s) can operate but with heavy reliance on temporary workarounds; system unstable; interruptions likely.	4 hrs
Severity 3	Critical: business function is operable but degraded; affected area/s of business can operate in a degraded state with reliance on temporary or permanent workarounds; system stable.	24 hrs
Severity 4	Non-critical: business function is operable with permanent workarounds.	48 hrs

Figure 19—Response Boundaries According to Severity Level.

TIP: Making Clear that Perfection Is Not Possible

As with any agreement involving software, it is very important to make plain the inherent limitations of the medium. Maintaining or servicing an automation system does not equate to delivering perfection. An example of a contract clause that makes this clear might be as follows:

> *Except to the extent contained in a separate, written agreement between Company and End User, Company's provision of services or Software under this Agreement does not create any new or different warranty. Company does not warrant that the Covered System will be error free or operate without interruption. Both parties understand that Systems and their many components have inherent limitations, and End User is ultimately responsible for determining whether any Covered System meets End User's requirements to achieve its intended results.*

The wisdom of contractually calling out this limitation is not confined to maintenance and service agreements—it is equally important when a system is being designed and installed in the first place.

Chapter 19

Legalities for Tough Economic Times

What is the role of the legal fine print when times are tough? Is it something to ignore because of the overhead of dealing with it (lawyers, negotiations)? Does it go on the back burner because the most important thing now—really, the only thing for heaven's sake—is your landing the project instead of the other guy?

I'm sure your company has its own answer for that and I am not here to judge it. Managing risk (including legalities) is one of the two or three most important buckets of institutional decision making that every automation enterprise must manage.

But I am here to urge thoughtful risk managers to consider the following thesis: some kinds of legalities deserve *increased* attention in an economic downturn. Why? Because project risk is magnified during tough times in at least three ways: the risk of insolvent partners, the likelihood of litigation and the risk of fewer projects.

The risk of insolvent partners. Let's face it, not everyone is going to come through an economic downturn alive. I can already hear the rationalizations in response: "We can weather this thing. We know what we're doing. I'm confident it's going to be the other company—not us—that ultimately falls down."

But does that response really work on automation projects? There are typically several parties involved with legal connections to your company. What happens if one of those project partners goes bust? Obviously, such a development could have a cascading negative effect, not just on the project as a whole, but on you. How well you do in such a meltdown, in my opinion at least, will depend on whether you gave attention to some very specific fine print.

First, for automation service providers:

- Are payments linked or unlinked? Unfair linkages include "pay when paid" clauses that can penalize your company for upstream problems that are not its fault. One potential upstream problem is an end user or middleman becoming insolvent.

- Liens, bonds, upstream retainage: These are third-party sources of money that can make your company whole even when a contracting partner goes belly up. If your company is involved with these things, do your project personnel have their eyes on all of the applicable requirements and notice deadlines?

For end users and owners, the issues are similar, only in reverse:

- Obviously, increased attention to detailed pre-qualification of all project participants, including subcontractors, is key. Require the balance sheets and have your controller scrutinize them.

- Is payment contingent on proof of lower tier payment (and is this condition observed)? It is one thing for the contract to require the submission of lien releases and waivers from underlying subcontractors; it is another to enforce these provisions strictly.

- Can teetering middlemen be bypassed? It is very important during difficult economic times to have the right to make payment to subcontractors and suppliers directly (or alternatively, to have the right to make payments by joint check) in the event of a middleman becoming financially unstable.

The likelihood of litigation. A second reality in a recession is that litigation is more likely. Companies will fight when money is short, long-term business relationships are less certain, and backs are against the wall. Are the negotiated rules of engagement for the war favorable? Does the battle have to be fought on your opponent's turf? Can you recover your expenses from your adversary if you win? Addressing those questions before problems start can be an investment with a huge return.

The risk of fewer projects. A third reality is that because projects are scarce, each one can assume an outsized importance. This means that there is much less company resilience in the face of the upside-down project—and no resilience whatsoever in the face of a project that is suffering a meltdown. For this reason, two questions must always be asked in an economic downturn:

- Is the project a mine field? Although it's tough to turn away business in tough economic times, some automation projects are clearly minefields, and you are betting on walking a perfect path to make it through unscathed.

- Is liability restricted? Limiting total liability to the contract price is fair. Excluding consequential damages should be a deal breaker if not permitted.

I know what you're thinking. The guy writing this is a lawyer who makes his living advising automation companies throughout North America and beyond—and who therefore has a vested interest in promoting the use of lawyers. Well, true, but that just leads us to another reality that unhappily circumscribes all of the others: there is no avoiding lawyers in a recession.

In other words, see us now or see us later.

CHECKLIST: Four Legal "Gut Checks" for Automation Providers during Tough Times

✔ **Is the middleman solvent?** If your company is contracting through a GC or other middleman, do you have any sense for that company's solvency? If there is any doubt, either a direct contractual relationship with the end user and/or some other third-party payment guarantee (e.g., a payment bond) should be explored.

✔ **Is liability restricted?** The upside of a smaller project is, by definition, smaller. Make sure that the downside is equally small. Limiting liability to the contract price is fair. Excluding consequential damages should be a deal breaker if not permitted.

✔ **Does your company have a lien strategy?** Contract managers should know and track the deadlines for lien recording in the project's jurisdiction.

✔ **Is the project a mine field?** Although it's tough to turn away business in tough economic times, some automation projects are clearly mine fields and you are betting on walking a perfect path to make it through unscathed. Trust your inner lawyer!

Chapter 20

Auditing Legal Health

Is it possible to conduct a self-audit to determine the legal health of a company in the automation industry? I would answer this question with a fairly confident—although very much qualified—"yes." The reason for the qualification is that the legal health of a business has as much to do with its culture and people (the moving parts, if you will) as it does with any "screen shot" that can be taken during an audit. Still, one difficult-to-dismiss screen shot that tells an important part of the story concerns the company's legal documents. Are they in order? Or, perhaps more relevant, do they even exist?

What do I mean by "legal documents"? I am taking a fairly liberal swipe at that. By legal documents I mean anything on paper that defines legal relationships—whether those relationships are with customers or between persons within an organization. What follows is a list of key legal documents that every company should have.

At the risk of being obvious, it is important for an automation company to have its own standardized **terms and conditions**, whether in traditional contract form or as an attachment to a proposal. Come to think of it, a standardized form of **proposal** is a good idea too—especially if there is some likelihood that the proposal will simply be accepted via the issuance of a purchase order by a customer (by the way, most companies should have their own standardized **purchase order** templates as well).

Does the company typically deal with demanding customers that transmit non-negotiable, one-sided terms? If so, then some sort of standard **contract rider** or addendum that asserts the terms that your company positively cannot afford to live without (the Control System Integrators Association has a good one that some of its members have used with great success—see Chapter 7).

An **employee handbook** that lists all employee rules and obligations at the company is an indispensable human relations tool for any sizeable American business. Is it important to keep employees from departing to a competitor or customer? Then consider a **non-compete agreement,** with a **confidentiality stipulation** along for the ride. The latter will protect company secrets from disclosure in the event that state laws do not.

Various **employee benefit documents** also ought to be generated, including retirement, pension, profit sharing and 401(k) and employee stock ownership plans, as well as health, dental, vision, disability, life and tuition reimbursement plans.

Do you remember why you decided to do business in the first place as XYZ Integration Services LLC as opposed to John Smith, Professional Engineer? Almost certainly it was because at some point you were advised, among other things, that the corporate form provided a shield against potential personal liability. But that shield may not be so sturdy after all if you do not keep corporation documents current, including the **corporate minute book, bylaws** and other important **corporate records.**

Unlike some other construction or installation enterprises, perhaps the most important assets of a company in the automation arena are its ideas. Are they protected? They are more likely to be if the company maintains an **intellectual property schedule** of its inventions, business methods, patents, trademarks and copyrights, has **research and development agreements** in place with its key employees, and has a **licensing agreement** for use with customers.

Have I exhausted the list? Not by a long shot. Any company of substantial size should have a **document retention policy** to determine whether to save or discard paper and electronic information. The company might also give strong consideration to formulating a **disaster recovery policy** for the unlikely (but devastating) possibility that a natural or man-made catastrophe disrupts business activity.

Finally, need I point out that having some sort of **succession plan** makes sense to the extent that the company wants to peacefully prepare for the time when the top persons are no longer around?

Is the list of "healthy company" documents (summarized fancifully below in Figure 20) comprehensive? Probably not. Are all of these documents equally important? Well, no. But you can imagine any one of the concerns addressed by these documents creating big problems if the documents in question are either nonexistent or hastily conceived. For that reason, among others, I leave it to you (with the help of your lawyer, of course) to sensibly prioritize them.

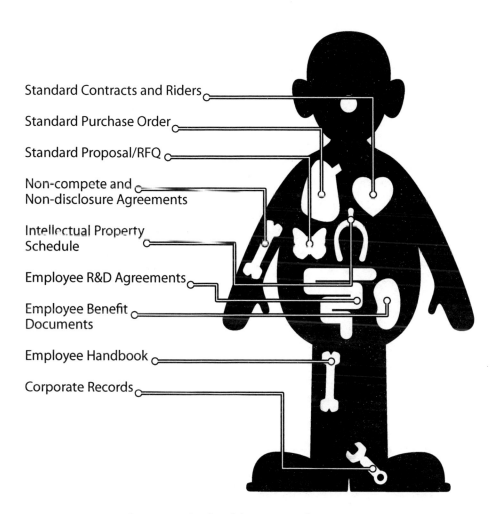

Figure 20—The "Healthy Company" Documents.

Chapter 21

Working with Attorneys

I have a confession to make: the goal of this book was never to educate you. It was to activate your "legal radar," so that you would recognize a legal matter when you saw one. That way you might be better equipped to know when to speak to your lawyer—and what to say when you do.

That is why I thought I would conclude this book by ghost-writing a memo to your attorney. That's right—ghost-write. That means I am writing a letter for you to send to your own attorney. (Satire alert: if you could see me now, you would note that my tongue is firmly planted within my cheek.)

Now, don't just clip this chapter out of this book; that would destroy the purpose. No, just re-type it, add your own two cents, and pretend it was from you. (For those of you who are concerned about intellectual property rights—which you should be if you paid attention during Chapter 9—the publisher and I hereby assign to you an irrevocable license to use the following words as if they were your own.)

* * *

To: My Attorney

From: One of Your Many Clients

Re: Our Relationship

Happy Belated New Year! I realize you are away on your annual vacation to Barbados this month, but I thought I would try to catch up with you with a few helpful suggestions for our business relationship. Please accept this for what it is—the equivalent of messages in a suggestion box. I am sure that you have a few things you would like to put in a suggestion box for me as well—

like the way our company literally continues to do business on a handshake (I know that drives you crazy) or that time we waited 180 days before paying you. Anyway, here are my six suggestions:

No. 1: **Use plain English.** Although I am an engineer and run a $10 million a year business, I am not a lawyer and it does not impress me when you speak in legal jargon. Instead, it makes me think that you are trying to bedazzle me with how smart you are. Translate your legal terminology and then I will think that you are truly smart. Don't say: "This term is probably unenforceable based on the doctrine of laches." Tell me: "There's a good chance the law will say they waited too long to hold you to this requirement and you can get out of it."

No. 2: **Sometimes I need a risk assessment, not a rewrite.** Don't always rewrite the contracts that I or my company send you. Sometimes, we are given contracts on a take it or leave it basis, so our choices are really down to two: signing the thing or walking away. It would be helpful for us if you could learn how to assess the legal risk in a page or so. That would allow us to either build the risk into our bid or decide that the risk is simply too high, and walk away.

No. 3: **Take the time to understand our business.** We are not the high tech equivalent of a plumber. We are integrators, manufacturers, engineers, automators and control system specialists. Know what a PLC is. Be able to distinguish MEP from SCADA. No, we are not going to ask you to do any programming anytime soon, but it would give us a bit more confidence that you understand our legal risks if you could at least describe what we do in a sentence.

No. 4: **Give thought to throwing us a freebie or two.** No, I'm not talking about tickets to an NFL game or a case of wine. I'm talking about an occasional off the clock phone call, or joining us for a brown bag lunch gathering of our project managers to talk about the ten worst contract clauses or how to protect the IP of our code. If you invest in us, we'll invest in you.

No. 5: **Do the math.** Don't always leave it to us to decide whether a lawsuit is worth it—we might be too irrational at the time to think about that. We would appreciate some input from your cooler head. If even a win would show up as a loss in our financial statements, tell us.

No. 6: **Give me your cell phone number.** This may seem irrational, but I would like to be able to reach you when I have a problem and I don't always succeed. Suffice it to say that your secretary and I are now on a first-name basis. Enough said.

CHECKLIST: 10 Additional Tips for Getting More Value Out of Your Lawyer

- ✔ **Don't be bashful about asking for billing rates.** Lawyers charge by the hour. You would not be bashful about asking for labor rates from a subcontractor or supplier. Lawyers are a type of subcontractor and you should know their rates too.

- ✔ **Sometimes the highest billing rate is actually cheaper.** The tradeoff between higher and lower rates is not always obvious. Indeed, on some types of tasks, the level of experience required to work efficiently makes a partner-level attorney less expensive than using an associate with a lower billing rate. Consider a partner with a billing rate of $400 per hour who spends 10 hours on a task versus an associate attorney with a billing rate of $250 per hour who spends 20 hours on the same task. The partner in this situation is the better deal.

- ✔ **There are alternatives to hourly billing.** It is not at all out of bounds to ask your lawyer, "How much is this going to cost?" Most lawyers are willing to provide an estimate (and if they are not, something is wrong). Depending on the project, they may or may not be willing to agree to a flat fee (of course, it goes without saying that flat fees are usually set at a level that are more profitable for the law firm than hourly rates). Another alternative that some lawyers are using is the concept of a success fee, which is a type of bonus that is awarded upon the occurrence of a favorable event.

- ✔ **Ask for an "hours-loaded" budget.** Most lawyers provide budgets for larger projects if asked (and the best ones, even if not asked) but the ideal, as you would suspect, is an hours-loaded budget. Such a budget reflects more precisely the assumptions being made and provides comfort in those circumstances (particularly involving unpredictable litigation) where a budget is blown.

- ✔ **What is the plan?** Hours-loaded budgets will also reveal whether your attorney has a plan for addressing the legal problem. Is there a logic and sequence to what the lawyer is proposing to do? Is the lawyer assuming control of events? Or is he or she merely reacting to what others are doing? What is the endgame?

✔ **Preventive medicine.** Automation companies that call their lawyers only in the event of lawsuits appear to have a strange—and inexplicable—love of legal fees. Time and money spent on using your lawyer for contract template creation, insurance program review, project risk assessment, insurance claims and training can avoid future (and potentially far more costly) problems.

✔ **All lawyers are not created equal.** If your instinct tells you that the storefront office lawyer whose kids play soccer with yours is not in the same business as the lawyer who took your deposition last year, trust your instinct. Although many states frown on lawyers labeling themselves as specialists in a particular area of law, it is not rocket science to realize that lawyers' skill and experience levels are not identical.

✔ **Copy me.** Your lawyer should send you copies of literally everything generated in the matter he or she is handling for you. It should not make a difference that you might not understand everything. You are paying for it. Insist on being copied on everything.

✔ **Do it yourself.** Why pay for a paralegal to gather and organize your project files at $185 per hour when you may have capable personnel available who could do that very same work for less? Tell your lawyer that you want your people to be deputized wherever possible.

✔ **Do the math.** Using a lawyer can be thought of as an investment. Will that investment pay a return—or at least prevent a larger loss? Your lawyer should be willing to give you some sense for the odds of success of any significant matter you are about to embark upon. Measuring the costs of that investment against the probable return is something you are doing in your business anyway. Don't hang up your business sense just because you are "lawyering up."

Glossary

Additional insured endorsement: An endorsement to an insurance policy that provides that a second project participant is also insured by the policy on some basis.

Administrative proceeding: A court-like trial or hearing conducted by a government agency. Often such a proceeding can be a prerequisite to obtaining relief in a court and is binding if not appealed or challenged.

Advocacy system: The system for dispensing justice in which the development of the proceedings is largely dependent upon the competitive activities of lawyers for the parties, with the court acting as umpire.

AGC: Associated General Contractors.

AIA: American Institute of Architects.

AIA forms: The preeminent set of U.S. contract forms for design and construction projects disseminated by the American Institute of Architects.

American Rule: The default rule for payment of attorneys fees in the U.S. legal system in which, regardless of winner, each side pays its own fees. The rule can change to "loser pays" if specified in a law or contract.

ANSI: American National Standards Institute.

Answer: The formal, written response to the allegations of a lawsuit in U.S. courts. Typically this is also the moment when the responding party asserts any claim that it may have against the party filing the original claim or other parties.

Appeal: A proceeding in which a higher court or other panel, often consisting of more than one person, is asked to reconsider a ruling by a court or other decision-making body.

Arbitration: A private, court-like proceeding in which parties agree to resolve their disputes as an alternative to a court of law. Contracts to arbitrate are binding, with the decisions by arbitrators (called "awards") final with no right of appeal.

Arbitrator: The person or persons rendering the award in an arbitration. Such persons can be lawyers, retired judges, experts or even laypersons.

Aspirational standard: A type of automation standard intended to be internal to the organization that is applying it and not enforceable by any outsider. Often these standards are "stretch goals" to which an organization aspires.

Assign or assignment: To transfer one's legal rights to another.

ATEX: European Union directive governing work in an explosive environment.

Award: The name of the document concluding an arbitration—the equivalent of the term "judgment" in a court of law. The term can also mean the decision to give a project to a particular bidder.

Backcharge: A reduction in the compensation that would otherwise be paid by one project participant to another due to amounts allegedly owed by the recipient to the payor.

Background patent: A patent that is a building block or foundation of a second patented invention.

Battle of the forms: A situation arising under the Uniform Commercial Code when two companies exchange competing terms and conditions for a particular project. When this happens, the UCC has rules for resolving inconsistencies between the two "forms."

Bench trial: A trial in which a judge decides the outcome of a case without the assistance of a jury.

Bond: A written instrument by which a surety guarantees the payment or the performance of a project participant.

Brownfield project: A project in a pre-existing industrial area, typically involving re-use of abandoned or under used property.

Builder's risk insurance: A type of property insurance designed to protect structures or production areas while they are being constructed or renovated.

Cardinal change: A change in a project of such magnitude (or the cumulative tipping point reached as a result of multiple changes) where the project no longer can be said to be of the scale or type generally contemplated at its start, justifying either the cancellation or renegotiation of the entire contract.

Case law: A written court decision, usually as a result of an appeal. See common law.

CCIP: Contractor-controlled insurance program.

Change: A modification of the previously-defined scope or duration of work for a project.

Change order: A written amendment to an existing project agreement, most commonly for the purpose of modifying a project's scope and/or duration.

Claim: Seeking relief in the form of additional money or time from another project participant.

Claimant: The person or company asserting a claim. The term is also used in the place of "plaintiff" in an arbitration or administrative proceeding.

Code standard: An automation standard with the force of law.

Commercial general liability (CGL) insurance: Standardized policy insuring against liability for bodily injury (or death) or property damage caused to others.

Common law: A type of law in which rules applied in previous court disputes are used to decide new disputes with similar fact patterns. Such written "case law" is constantly evolving, growing organically over years and decades, with the rate of development dependent on the happenstance of new cases. Compare with statutes and regulations. Common law countries include the U.S. and U.K.

Complaint: One of the documents, in addition to a summons, initiating a lawsuit in the U.S. A complaint alleges a right to judicial relief as a result of the actions or inactions of one or more defendants.

Confidentiality agreement: See non-disclosure agreement.

ConsensusDocs: A set of contract forms disseminated by a disparate group of trade associations for use in design and construction projects. The forms purport to be more fair than other forms in that they do not favor any particular constituency in a project.

Consequential damages: Lost revenues and other monetary losses resulting from an event giving rise to a claim. Consequential damages do not include the cost of implementing, repairing or restoring the functionality of a system (which are known as direct damages).

Construction manager: A project participant hired to implement a project, either as an advisor to the end user or serving (on the end user's behalf) as the "at risk" prime contractor. In either form the construction manager works closely with the design team in an effort to blend construction expertise with design expertise at an early stage.

Contract: An agreement that can be enforced in a legal proceeding.

Contractor-controlled insurance program: Wrap-up insurance coverage that is purchased and controlled by a contractor.

Contractual liability exclusion: An insurance policy exclusion that, generally speaking, attempts to exclude business risks assumed in contracts that go beyond the basic type of risk the insurance policy was intended to cover. This does not mean the exclusion precludes coverage of all losses caused by an insured while performing work under a contract.

Cost plus agreement: A contract in which the agreed compensation to the designer and/or installer of an automation system is the out-of-pocket cost of producing it, even if not yet known, plus an agreed fee (profit), sometimes expressed as a percentage of the cost.

Coverage grant: The set of "big picture" sentences in an insurance policy (often at the beginning, but not always) that lay out the overall intent of the policy and which confers coverage upon the policyholder.

CSIA: Control System Integrators Association.

Defendant: A target of a claim in a legal proceeding.

Deliverable: Materials, equipment and/or services to be provided in connection with a project.

Design-bid-build: A project delivery method in which an owner or end user first hires a designer to create the design of a system, then calls upon a separate project participant to bid the cost of implementing that design.

Design-build: A project delivery method in which the owner or end user hires a single project participant to provide both the design of a system and its implementation.

Design-build-operate-maintain: A type of design-build project in which the design-builder commits to a period of operating and/or maintaining a system or facility for the end user following completion.

Design-builder: A project participant hired to provide both the design of a system and its implementation.

Design spec: A specification calling for the implementation of a system with particular static characteristics.

Direct damages: The cost of implementing, repairing or restoring the functionality of a system. Direct damages do not include lost revenues and other monetary losses resulting from an event giving rise to a claim (which are known as consequential damages).

Disclaimer: A contract clause that expressly rejects or renounces liability for a particular type of risk.

Dispute resolution clause: One or more contract provisions laying out a method for resolving disputes on a project, often mandating informal or formal attempts at conciliation as a condition to seeking a decision in a legal proceeding.

Downstream: Contract relationships that are more remote from the prime contract relationship than the one in question.

Dragnet clause: A sweeping contract term that strives to resolve any lack of clarity in legal risk against a particular project participant. The literal translation of such clauses is "whenever there is uncertainty, it goes against you."

Employee handbook: A manual containing policies applicable to the employees of an organization.

Employee R&D agreement: An agreement by which an employee agrees to assign to a company his or her ownership rights to any invention created in connection with the employee's work for the company.

Endorsement: An amendment to an insurance policy typically changing the duration or scope of coverage.

Energy Star: An international standard for energy efficiency that began in the United States but is now used in other countries as well.

EJCDC forms: A set of contract forms disseminated by the Engineers Joint Contract Documents Committee for use in design and construction projects.

Engineer-procure-construct or EPC project: A type of design-build used in process facilities, often on a "turnkey" basis, in which the predominant deliverable is a particular output or performance level.

Exception: A "carve out" within an insurance policy exclusion that adds back coverage.

Exclusion: Provision within an insurance policy that subtracts coverage.

Expert witness: A witness in a legal proceeding with specialized knowledge, experience or training that is helpful for the resolution of a particular case. Witnesses confirmed as such may be permitted to offer opinions that can be relied upon by a judge, jury or arbitrator.

False claim act: Laws that create liability for asserting a false or exaggerated claim against a governmental entity.

FIDIC forms: The preeminent set of international contract forms disseminated by the Fédération Internationale Des Ingénieurs-Conseils (International Federation of Consulting Engineers) for use in design and construction projects.

Force majeure: A major, unanticipated event or other "act of God" that has the legal effect of relieving a project participant from some or all of its contract obligations. The list of matters triggering this relief (e.g., hurricane, fire, terrorism) is often negotiated.

Foreclosure: A lawsuit to enforce a mechanic's lien. Literally, such suits seek a court order selling the real estate encumbered by the lien to pay off the amount owed.

Green Globes: An environmental rating and evaluation system propagated by the Green Building Initiative.

Greenfield project: A project on previously undeveloped land.

Guarantee or guaranty: A contract to pay the debt of another if the principal debtor is unable to make payment.

HACCP: Hazard analysis and critical control points.

Hold harmless clause: A contract term in which a project participant agrees not to make a claim against another arising out of a particular type of risk.

IEEE: Institute of Electrical and Electronics Engineers.

Implied warranty: A type of warranty sometimes imposed by the law, even if never promised outright in a contract or other offer. Two such warranties of particular significance to automation projects are (1) the implied warranty of merchantability and (2) the implied warranty of fitness for a particular purpose. It is possible to remove the application of such warranties through a proper disclaimer.

Implied warranty of fitness for a particular purpose: A type of warranty imposed by the law in the situation when a seller with relevant expertise has reason to know of a buyer's particularized need and the buyer reasonably relies upon that seller's expertise in providing a product to meet that need.

Implied warranty of merchantability: A type of warranty imposed by the law in many sales situations—a warranty in which a seller promises that a buyer's most basic expectations regarding a product will be met (i.e., the type of expectations any ordinary buyer would have with any such product). A common way to avoid this warranty is to specify that a product is being provided "as is."

Incorporated document: An extraneous document that is specifically identified or described in the contract and therefore included as a binding obligation.

Incorporation by reference: The identification within a contract of a extraneous document for the purpose of adding the obligations in that document to those of the contract.

Indemnity or indemnification: An agreement to compensate another project participant in whole or in part in the event of claims asserted by third parties.

Industrial exemption: The deep-rooted, but increasingly challenged, rule that skilled persons working within industrial facilities are exempt from professional licensing laws.

Industry standard: A type of automation standard disseminated by industry organizations without the force of law yet recognized as carrying weight within the profession.

Integrated project delivery: An experimental project delivery method in which the owner or end user signs a single agreement with multiple companies at the earliest possible stage for the purpose of collectively designing and implementing a system or systems. IPD differs from design-build in that a nucleus of critical participants, sometimes including key subcontractors to the design-builder, are elevated to equal partners in the design and construction process.

Integration clause: A contract clause precluding the enforceability of obligations outside of those specified in the contract itself and any incorporated documents. A typical target of such clauses is communications preceding the signing of the contract.

Intellectual property: Ideas that can be owned.

Intellectual property schedule: A list of all intellectual property owned by an organization.

ISA: International Society of Automation.

Joint check: A check payable jointly to two payees. Because in theory neither payee can cash the check without the endorsement of the other, such checks are used as a means of ensuring payment to both.

Jury trial: A form of court proceeding in which laypersons collectively determine the facts of a case, as revealed to them through the advocacy system, then apply the law to those facts under the guidance of a judge.

LEED: Leadership in Energy and Environmental Design, a third-party green certification program administered by the United States Green Building Counsel (USGBC).

Letter of intent: Temporary agreement between parties that are in the process of negotiating a more complete deal. Such a letter, by definition, is non-binding in the sense that there is no guarantee the ultimate "full deal" will be agreed upon—although it does bind the parties to negotiate in good faith and to live up to the terms of the Letter of Intent itself.

License: In a general sense, the term means permission. With regard to professionals (e.g., engineers), it means the government has granted the licensee permission to operate within the professional field in question. With regard to intellectual property, it means the owner of intellectual property has granted the licensee permission to use the property to the extent permitted by the license.

Limitation of liability clause: A contract provision that caps the liability of a project participant to a particular maximum amount.

Liquidated damages: An agreement between project participants that if a particular adverse event occurs, the victimized party will not be required to prove that it suffered any particular amount of loss but instead will be entitled to collect a pre-defined amount of money from the other party. Such damages typically are not enforceable if there is no actual harm to the victim such that the amount to be collected is only a penalty.

Lump sum agreement: A contract in which the total amount to be charged for a particular scope is defined and does not vary, even if the actual cost incurred is higher or lower.

Maintenance and service agreement: A contract to sustain and/or refurbish an automation system on some continuing basis.

Mechanic's lien: The right of an unpaid project participant (mechanic) to record a claim in the land records against real estate that has been improved by its work. Such a lien literally entitles the contractor to force the sale of the real estate and improvements to pay the amount owed. Although often pursued at the same time as the contractor's right to seek payment under the agreement by which it was hired, the recording of such a mechanic's lien is technically separate and may be against a different party altogether.

Mediation: A non-binding process in which a third-party neutral attempts to broker the settlement of a dispute between two or more parties. Under most circumstances, the discussions are protected by law from being used for or against any of the participants.

Motion: A written request for action by a court or other person with adjudicative authority. Motions are how most U.S. legal proceedings move from Point A to Point B.

MRO: Standing for "maintenance, repair and operations," the term primarily refers to ongoing activities to keep an industrial facility and its systems efficiently running. However, the term sometimes is more expansively used by some organizations to describe even new capital projects within the industrial sphere.

NEC: National Electrical Code.

Negligence: The breach of a legally-recognized duty that results in injury to the party to whom the duty was owed.

NFPA: National Fire Protection Association.

No damages for delay clause: A contract term in which a project participant agrees that its sole remedy in the event of a project delay will be an extension of time and not additional money. Such terms are enforceable in most jurisdictions in the absence of exceptional circumstances.

Non-compete agreement: A contract, typically between an employer and key employee, in which the employee, once his or her employment ends, agrees for a limited period of time not to compete with the employer within a particular sphere and territory.

Nondisclosure agreement or NDA: A contract in which one or both parties agree to keep confidential certain information learned from the other while doing business together. Such agreements typically are used in automation projects to protect against the disclosure of trade secrets, sensitive proprietary matter and/or intellectual property intentionally or unintentionally obtained during the course of designing or implementing a system. Another typical use of such contracts is during the "due diligence" (investigative) period preceding any significant business acquisition.

OCIP: Owner-controlled insurance program.

Order of priority clause: A contract term listing the order in which contract documents or parts of contract documents take precedence in the event of conflict between them.

OSHA: Occupational Safety and Health Administration.

Owner-controlled insurance program: Wrap-up insurance coverage that is purchased and controlled by an owner or end user.

Patent: A patent is a right to an idea that is granted to an inventor by the federal government. It generally lasts 20 years.

Patent license: The right to use an idea patented by someone else.

Pay-if-paid clause: A contract term providing that a subcontractor or subconsultant is only due payment to the extent the upstream contractor itself is paid. In theory, if the upstream contractor is never paid, the subcontractor or subconsultant is never owed.

Pay-when-paid clause: A contract term providing that a subcontractor or subconsultant is not due payment until the upstream contractor is first paid. In a number of jurisdictions such a term may be ignored if the time for payment has been unreasonably delayed.

Payment bond: A written instrument by which a third party called a surety guarantees that a contractor will pay its subcontractors and other vendors. If the contractor fails to do so, the surety can be liable to the subcontractors and vendors for the payment.

Performance bond: A written instrument by which a third party called a surety guarantees that a contractor will follow through with its contract obligations. If the contractor fails to do so, the surety can be liable to the party to which those obligations are owed.

Performance spec: A specification that calls for the creation of a system with particular dynamic or functional characteristics.

PIP: Process Industry Practices.

Plaintiff: A person or company asserting a claim in a lawsuit.

Pollution coverage: Insurance covering the release of chemicals within a process facility. A limited form of such coverage is contained within standard CGL policies.

Products and completed operations coverage: The extension of CGL insurance coverage to a period following a project's completion. This coverage is standard with most CGL policies.

Professional liability insurance: Known as errors and omissions, or E&O coverage, PL insurance is focused on a particular cause: negligence. For such coverage to respond there must be a breach of a professional duty resulting in a loss to others.

Project delivery method: An arrangement for allocating legal responsibility among a given set of participants in a project.

Prompt payment law: A law requiring the payment of downstream project participants within a prescribed period of time.

Proposal: An offer to provide materials, equipment and/or services, typically, but not always, in response to a request by a customer. Once accepted by a purchase order, a proposal and purchase order may together form a binding contract.

Purchase order: A legal instrument by which a project participant accepts the offer or proposal of another. A purchase order, together with other exchanged documents or performance by project participants, may form a binding contract.

Regulation: A rule adopted by an administrative agency.

Reservation of rights letter: A letter from an insurance company in response to a claim in which it agrees to defend its insured against the claim while not giving up its right to assert that the claim is not covered by the insurance policy in question.

Respondent: A term used in the place of "defendant" in an arbitration or administrative proceeding.

Response boundaries: In a maintenance and service agreement, the defined time periods in which an automation provider agrees to take action to address certain problems.

Retainage: An amount held back by one project participant from payment owed to another, typically 5-10%, as a means of ensuring completion of performance and correction of mistakes.

RFP: Request for a proposal.

RFQ: Request for a quote.

Rider: An addendum to an agreement.

Royalty: Compensation paid for the right to use someone else's intellectual property.

Scope of work: The equipment, material and services that a project participant has agreed to provide.

Security agreement: A sale of equipment or material in which the right of the seller to payment is protected by the seller's ability to reclaim the equipment or material in the event payment is not made.

Spearin doctrine: A legal rule that any specification imposed by an owner or end user upon a contractor carries with it the implied warranty that if the specification is followed, the result will be adequate.

Spec standard: An automation standard that must be fulfilled because of its incorporation within a contract.

Specifications or Specs: Anything in writing that purports to describe in itemized detail the work being done.

Statute: A law passed by a legislative body.

Summons: One of the documents, in addition to a complaint, initiating a lawsuit in the U.S. A summons literally notifies a defendant that it has been sued (and is now being "summoned" to court to respond to the allegations).

Surety: A company in the business of selling bonds that guarantee the performance of others.

Ts and Cs: Terms and conditions.

Termination for cause: A contract provision permitting the termination of the contract by one or more parties in certain prescribed circumstances.

Termination for convenience: A contract provision permitting the termination of the contract by one or more parties for any reason.

Terms and conditions: An imprecise term used by automation companies and their customers to describe "standard" legal language purportedly applying to a project. Such language can appear in RFPs and RFQs, proposals, purchase orders, contracts and online. The enforceability of such provisions depends on whether such terms were successfully incorporated within a larger contract and the existence of contrary terms.

Tort: Legal wrongs giving the injured party the right to make a claim against the wrongdoer. Negligence is a type of tort.

Turn-key: A term implying that a system is to be designed and/or constructed to the point of operability with minimal effort by the end user and/or owner, such that the operator need only "turn the key" the system will function.

UCC: Uniform Commercial Code.

UL: Underwriters Laboratories.

Uniform Commercial Code: Standardized laws on the books in U.S. states that govern a number of important types of business relationships. Of most significance to automation companies is the UCC's "Article 2," adopted in all but one state, which essentially writes a contract for project participants under certain circumstances when they do not agree upon a contract of their own or leave out an important term. The UCC heavily influenced its international equivalent, the United Nations Convention on Contracts for the International Sale of Goods (CISG).

Upstream: Contract relationships that are in greater proximity to the prime contract relationship than the one in question.

Waiver of consequential damages: A contract provision in which project participants agree not to seek consequential damages from each other in the event of a claim between them.

Waiver of subrogation: A contract provision in which a project participant agrees that its insurance carrier, after paying on a loss, will not be permitted to seek restitution against the ultimate wrongdoer. The purpose of such a provision is to avoid the hassle of further litigation that can result when subrogated insurers "stand in the shoes" of their policyholders after paying on a claim.

Warranty: A contract provision (or a promise implied by the law) in which a project participant assures another that certain facts are true or will happen such that the other party may seek relief if it turns out not to be true. Warranties can be written into a contract or imposed by the law. When offered outright, they are called express warranties. When imposed by the law, they are called implied warranties.

Workers' compensation: U.S. laws (with many international equivalents) governing work place injuries. Such laws impose a trade-off on both employers and employees, with employees giving up the right to seek damages from employers in return for wage replacement and medical benefits, and with employers released from negligence liability in return for being required to purchase insurance to cover the potential obligation.

Wrap-up insurance: An approach to risk in which one project participant purchases certain types of insurance for itself and all other key participants. The types of coverage typically addressed in a wrap-up include CGL, worker's compensation and property insurance, with other coverages (such as professional liability) left for the individual companies.

Index

accord and satisfaction 78
account stated 78
additional insured 35–36, 109, 111, 139
administrative proceeding 90, 139, 141, 151
advocacy system 82, 139, 147
AGC 17, 139
AIA 14, 17, 139
American Institute of Architects (AIA) 14, 17, 139
American rule 82, 139
ANSI 61, 139
answer 103, 124, 139
appeal 6, 40–41, 90, 139
arbitration 22, 44, 82, 89–90, 140
arbitrator 19, 44, 88–90, 140
aspirational standard 62, 64, 140
assignment 140
Associated General Contractors (AGC) 17, 139
ATEX 63, 140
attorneys 15, 82, 86, 121, 135
    fees 91, 98, 139
award 44, 137, 140

backcharge 29, 140
background patent 56, 140
battle of the forms 37, 140
bench trial 90, 140
bond 3, 35, 91, 113–115, 117–120, 128, 140, 152
brownfield 141
builder's risk insurance 141

cardinal change 4, 79–80, 141
case law 141
CCIP 111, 141
CGL 105–111, 141, 150
change order 2–3, 78, 80, 141
claim 3, 27, 29–30, 35, 43, 56, 78, 80, 86, 91, 95–96, 99–101, 103, 105–106, 108, 110, 112–114, 116–120, 138, 141
    false 93, 145
claimant 116, 120, 141
code standard 62, 64, 141
commercial general liability insurance 105
commissioning 11, 14, 17, 48, 88
common law 96, 99–100, 141
complaint 142, 152

confidentiality agreement 142
ConsensusDocs 142
consequential damages 3, 26, 31, 45, 73, 98, 129, 142–143, 153
construction manager 111, 142
contract 1–7, 9, 11–13, 15–17, 19–23, 25–31, 33–43, 50–51, 54, 57, 61–64, 66, 72–73, 76–77, 79–80, 82–83, 88, 90–93, 95–97, 99–101, 106–108, 111–112, 115, 117, 120–123, 125, 128–129, 131, 136, 138, 142
contractor-controlled insurance program (CCIP) 111, 141–142
contractual liability exclusion 108, 142
control system 2, 4, 15, 26, 28, 32, 35, 47, 56, 61, 65–67, 69–71, 79, 97–99, 124, 136
cost plus agreement 142
coverage grant 104, 143
CSIA 41, 43, 59–60, 62, 143

damages 22, 26–27, 37, 55–56, 73, 98, 105, 108, 154
    consequential 3, 26, 31, 45, 73, 98, 129, 142–143, 153
    direct 26, 142–143
    liquidated 2–3, 27, 148
defendant 142–143, 151–152
deliverable 11–12, 17, 27, 109, 143
design spec 5, 47–50, 143
design-bid-build 10–11, 143
design-build 10–11, 14, 143
design-build-operate-maintain 11, 143
direct damages 26, 142–143
disclaimer 37, 143
discovery 76, 82–84, 87
dispute resolution 3, 22, 29, 81, 143
dragnet clause 3, 23, 33, 144
duty 88, 95–96, 99–100, 107, 112, 149–150

EJCDC 17, 144
e-mail 75–77, 83, 86
employee handbook 131, 144
employee R&D agreement 144
endorsement 35–36, 104, 109, 111, 139, 144
Energy Star 70–71, 73, 144
engineer-procure-construct (EPC) 10, 144
EPC 10, 144

155

The Automation Legal Reference

exception 6, 10, 21, 40, 79, 104, 106–107, 144
exclusion 104, 106–108, 123–124, 144
expert 84, 86, 88–89, 91, 97, 140, 144

factory acceptance testing (FAT) 38
false claim 93, 145
FAT 38
fault 30–31
FIDIC 145
force majeure 3, 123, 145
foreclosure 116, 145
free from defects 33, 48

Green Building Initiative 70, 145
green globes 70–73, 145
greenfield 1, 145
guarantee or guaranty 35, 117–120, 145

HACCP 63, 145
hold harmless 145

IEEE 59, 63, 145
implied warranty 50, 145
    of fitness for a particular purpose 145–146
    of merchantability 145–146
incorporation by reference 3, 20, 146
indemnity or indemnification 2–3, 30–32, 45, 100, 108, 112, 146
industrial exemption 65–67, 146
industry standard 33, 61–62, 64, 146
infringement 15, 32, 44, 54–56
insurance 12, 26, 30, 32, 35, 45, 99, 103–110, 112, 117, 119, 138
integrated project delivery 11, 146
integration clause 20, 146
intellectual property 4–6, 13, 15, 17, 27, 31, 44, 53–54, 135, 147
    schedule 132, 147
International Society of Automation (ISA) 61, 147
invention 6, 27, 53–57, 132
ISA 59–61, 67, 97, 147

joint check 120, 128, 147
jury trial 90, 147

lawsuit 15, 54, 66, 72–73, 76, 81–82, 84–87, 89–90, 95, 105–106, 116–117, 136, 138
lawyer 2, 6, 10, 19–20, 39–42, 47–49, 53, 59, 61, 73, 75–77, 81–82, 89, 91, 96–97, 105–106, 112, 127, 129, 133, 135, 137–138
LEED 63, 70–74, 147

legacy 6, 9, 17, 27
letter of intent 42, 147
liability 16, 35, 37, 40, 42–44, 49, 73–74, 93, 95–96, 99, 105–109, 111, 122–123, 129, 132
license 6, 27, 44, 53–54, 56–58, 66, 68, 135, 147
licensing 56, 65–68, 132
lien 3, 91, 113–117, 120, 128–129
limitation of liability 3, 31, 45, 122, 147
liquidated damages 2–3, 27, 148
litigation 78, 81–82, 84, 87–88, 90–91, 98, 112, 127–128, 137, 153
lump sum agreement 148

maintenance 1, 17, 38, 58, 121–122, 124
maintenance and service agreement 121, 123–125, 148
mechanic's lien 3, 30, 82, 114, 116, 118, 148
mediation 87, 90, 148
motion 84–87, 148
MRO 1, 148

National Electrical Code (NEC) 148
National Fire Protection Association (NFPA) 62, 149
NDA 149
NEC 59, 148
negligence 30–31, 43, 95–97, 99–101, 107–108, 110, 149
negligent 32, 99
NFPA 59–60, 62, 149
no damages for delay 27, 149
non-compete agreement 132, 149
nondisclosure agreement 57, 149
notice 5, 22, 29, 36, 55, 57, 76, 79–80, 108, 116, 120, 128

OCIP 111, 149
order of priority 3, 92, 149
OSHA 59, 63, 100, 149
owner-controlled insurance program (OCIP) 111, 149

patent 15, 32, 44, 47, 53–56, 58, 132, 140, 149
pay-if-paid 150
payment bond 114, 117–118, 129, 150
pay-when-paid 3, 150
performance bond 117–119, 150
performance spec 5, 28, 47–50, 72, 150
PIP 59, 63, 150
plaintiff 141, 150
PLC 15, 47, 59, 61, 115, 136

pollution coverage 109, 111, 150
products and completed operations coverage 109, 150
professional liability insurance 12, 107, 150
project delivery method 9–12, 143, 146, 151
prompt payment law 120, 151
proposal 2, 7, 13, 16, 19–20, 23, 38, 40–41, 45, 47–48, 95, 131, 151
purchase order 2, 9, 13, 15–16, 19–20, 39, 50, 72, 95, 111, 131, 151, 153

regulation 23, 61, 66–67, 96, 100, 151
reservation of rights letter 151
respondent 151
response boundaries 123–125, 151
retainage 3, 34, 128, 151
RFP 20, 151, 153
RFQ 16, 23, 40, 151, 153
rider 7, 41, 43–45, 131, 151
royalty 54, 58, 152

safety 63, 65, 99–101
SAT 38
scope of work 3, 16–17, 19, 21, 23, 28, 73, 152
security agreement 120, 152
site acceptance testing (SAT) 38, 97
software 1, 4, 6, 9, 14, 17, 19, 26–27, 33, 43–44, 54, 56–58, 60, 66, 72, 79, 114, 123, 125
solicitation 45
Spearin doctrine 50, 152
spec standard 61–62, 64, 152
specification or spec 4–5, 16, 19, 21, 28, 38–39, 43, 47–51, 54, 63–64, 76, 92–93, 152
standard 23, 33, 38, 48, 51, 59–64, 70, 72, 97, 99, 101, 140–141, 144, 146, 152
state of the art 16, 48
statute 67, 152
summons 152
surety 35, 114, 117–119, 152

termination for cause 152
termination for convenience 29, 152
terms and conditions 4, 13, 16, 25, 30, 38–41, 79, 95, 131, 152
tort 153
Ts and Cs 152
turn-key 10–11, 153

UCC 140, 153
UL 60, 63, 153
Uniform Commercial Code (UCC) 14, 140, 153
USGBC 70–71, 147

waiver of consequential damages 26, 153
waiver of subrogation clause 112, 153
warranty 3, 14, 16, 36–37, 43, 48, 50–51, 73, 106, 122, 124–125, 154
workers' compensation 110–111, 154
wrap-up insurance 142, 149, 154

About the Author

Mark Voigtmann is a partner in Faegre Baker Daniels, an international law firm with offices in the United States, China, and the U.K. He has advised automation providers, OEMs and their end user customers in numerous matters in North America and beyond, from the structuring of projects and negotiation of contracts to the resolution of insurance coverage questions and other complex automation disputes. He lives in Indianapolis.